Adaptive Wavelet Schwarz Methods for Nonlinear Elliptic Partial Differential Equations

Dominik Lellek

Logos Verlag Berlin

λογος

Bibliografische Information der Deutschen Nationalbibliothek

Die Deutsche Nationalbibliothek verzeichnet diese Publikation in der
Deutschen Nationalbibliografie; detaillierte bibliografische Daten sind
im Internet über http://dnb.d-nb.de abrufbar.

ISBN 978-3-8325-4067-8

Logos Verlag Berlin GmbH
Comeniushof, Gubener Str. 47,
D-10243 Berlin
Germany

Tel.:   +49 (0)30 / 42 85 10 90
Fax: +49 (0)30 / 42 85 10 92
http://www.logos-verlag.de

Adaptive Wavelet Schwarz Methods for Nonlinear Elliptic Partial Differential Equations

Dissertation

zur
Erlangung des Doktorgrades
der Naturwissenschaften
(Dr. rer. nat.)

vorgelegt dem

Fachbereich Mathematik und Informatik
der Philipps-Universität Marburg

von
Dipl.-Math. Dominik Lellek

geboren am 03. Januar 1986
in Wetzlar

Am Fachbereich Mathematik und Informatik
der Philipps–Universität Marburg als Dissertation
eingereicht am: 05. Februar 2015

Hochschulkennziffer: 1180

Erstgutachter:
Prof. Dr. Stephan Dahlke, Philipps–Universität Marburg

Zweitgutachter:
Jun.-Prof. Dr. Thorsten Raasch, Johannes Gutenberg–Universität Mainz

Tag der mündlichen Prüfung: 05. Mai 2015

Acknowledgements

I start with expressing my sincere gratitude to my advisor, Professor Stephan Dahlke, for giving me the opportunity to become a member of his group. I would like thank him for his support and for valuable discussions and suggestions. By encouraging me to attend numerous conferences and workshops, he made it possible for me to get in touch with a large number of researchers working in numerical analysis and to broaden my mathematical horizon.

Furthermore, I would like to thank Junior Professor Thorsten Raasch for his willingness to write the second referee report.

I would also like to thank Professors Shaun Lui and Rob Stevenson for a number of helpful discussions and their very friendly cooperation.

Moreover, I am grateful to the former and current members of the workgroup Numerics and Optimization in Marburg for being fantastic colleagues. Thank you for all the mathematical and non-mathematical discussions, numerous more or less successful games of *Doppelkopf* and for the open and friendly atmosphere in general. In particular, I would like to thank Gregor Kriwet for being an incredibly tolerant and pleasant office mate and Jens Kappei for many helpful discussions in early stages of this thesis, and for being a very nice flatmate.

I am especially indebted to my friends for keeping me distracted from mathematics whenever necessary. Thank you all for being there and for having made the time in Marburg a valuable experience.

Last but not least, I would like to thank my parents for their constant support and encouragement.

Zusammenfassung

In dieser Arbeit beschäftigen wir uns mit der Konstruktion numerischer Verfahren zur Lösung nichtlinearer elliptischer partieller Differentialgleichungen. Solche Gleichungen finden vielerlei Anwendungen in den Naturwissenschaften, sind aber nur in Ausnahmefällen explizit lösbar. Daher besteht ein Bedarf an geeigneten numerischen Verfahren zur Approximation der Lösung.

Gängige Methoden zur numerischen Behandlung sowohl linearer als auch nichtlinearer elliptischer Gleichungen sind *uniforme* Verfahren basierend auf einer Diskretisierung mit Finiten Elementen. Die Grundidee solcher Verfahren ist es, die Lösung durch Funktionen zu approximieren, die mithilfe eines Gitters mit gleichmäßiger Weite konstruiert sind. In vielen Anwendungen, etwa auf polygonalen Gebieten, haben die Lösungen jedoch lokale Singularitäten, die es sinnvoll erscheinen lassen, *adaptiv* zu arbeiten. Darunter ist anschaulich zu verstehen, dass besonders dort viele Freiheitsgrade bzw. eine hohe Auflösung gewählt werden, wo die Lösung Singularitäten zeigt. Dort, wo sie glatt ist, kann auch mit weniger Aufwand eine gute Approximation erreicht werden. Diese Verfeinerungsstragien sollen ohne a-priori-Informationen über die Gestalt der unbekannten Lösung auskommen, sondern die Verfeinerung anhand von im Laufe des Verfahrens gewonnener Informationen vornehmen. In der Praxis werden adaptive Finite-Element-Methoden häufig eingesetzt. Der theoretische Konvergenzbeweis für solche Verfahren war jedoch lange offen, inzwischen sind zumindest in einigen Fällen Konvergenzraten beweisbar, siehe etwa [9, 91, 106].

Ein alternativer, vielversprechender Ansatz zur Konstruktion von adaptiven Verfahren zur numerischen Lösung von Operatorgleichungen ist die Verwendung von *Wavelets*. Diese Funktionen bilden eine Basis eines Funktionenraums, die im Wesentlichen aus dilatierten und translatierten Versionen einer einzelnen oder sehr weniger Funktionen besteht. Methoden basierend auf Wavelets wurden, beginnend in den 1980er-Jahren, zunächst insbesondere in der Signalverarbeitung und Zeit-Frequenz-Analyse eingesetzt. Für einen Überblick über Waveletkonstruktionen und Anwendungen, siehe beispielsweise [20, 48, 90, 125]. Insbesondere gibt es Waveletkonstruktionen mit stückweise glatten, lokalisierten Funktionen, welche verschwindende Momente besitzen und klassische Glattheitsräume wie *Sobolev-* oder *Besovräume* charakterisieren, siehe etwa [29, 45, 95]. Diese Wavelets können zur Diskretisierung von Operatorgleichungen verwendet werden und erlauben es, mit wenigen Koeffizienten eine gute Approximation an die Lösung zu erreichen, auch wenn diese Singularitäten aufweist. Diese Eigenschaften wurden in der grundlegenden Arbeit [25] ausgenutzt, um ein adaptives Wavelet-Verfahren zur Lösung linearer elliptischer Operatorgleichungen zu entwickeln, das nicht nur beweisbar konvergent ist, sondern unter sehr allgemeinen Voraussetzungen auch *asymptotisch optimal*. Darunter versteht man, dass das Verfahren die Konvergenzrate der *besten N-Term-Approximation* reproduziert, also der besten Approximation mit N Freiheitsgraden in der gegebenen Wavelet-Basis, gegebenenfalls eingeschränkt auf sogenannte baumstrukuturierte Indexmengen. Dank der Charakterisierung von Glattheitsräumen durch Wavelets kann diese Rate mithilfe der Regularität der Lösung in einer bestimmten Skala von Besov-Räumen beschrieben werden. In vielen Fällen, so etwa auf polygonalen Gebieten, ist die Regularität in dieser Skala höher als die Sobolev-Regularität, welche die Konvergenzrate klassischer uniformer Verfahren bestimmt. Daher bieten in solchen Situationen adaptive Wavelet-Verfahren einen theoretisch nachweisbaren, aber auch in Experimenten beobachtbaren Vorteil gegenüber uniformen Verfahren. In der Folgezeit wurde die Anwendung solcher Verfahren auf weitere Problemklassen ausgeweitet. Als Beispiel seien indefinite Probleme wie Sattelpunktprobleme, siehe etwa [26, 31, 36], und Integralgleichungen, siehe [43, 70] genannt. Insbesondere wurden in [27] konvergente und asymptotisch optimale adaptive Wavelet-Verfahren für eine Klasse nichtlinearer elliptischer Probleme entwickelt.

Eine zentrale Voraussetzung für die Anwendung solcher adaptiver Wavelet-Verfahren ist die Verfügbarkeit einer geeigneten Wavelet-Basis. Auf dem Einheitskubus oder affinen Bilder desselben gibt es eine Reihe möglicher Konstruktionen. Für allgemeinere Gebiete wie nicht-konvexe polygonale Gebiete sind zwar verschiedene Techniken zur Verallgemeinerung solcher Konstruktionen bekannt, siehe etwa die Konstruktion mittels nichtüberlappender Gebietszerlegung aus [46]. Diese können jedoch zu ungünstigeren Konditionszahlen oder einer schwierigeren Implementierung führen, da beispielsweise bei einer nichtüberlappenden Gebietszerlegung die Wavelets an der Rändern der Teilgebiete geeignet zusammengefügt werden müssen. Diese Probleme können in vielen Fällen vermieden werden, wenn anstatt von Wavelet-Basen aggregierte Wavelet-Frames verwendet werden. Dieser in [104] entwickelte Ansatz beruht auf einer überlappenden Gebietszerlegung in einfachere Teilgebiete, die etwa affine Bilder des Einheitskubus sind. Auf den Teilgebieten können dann bekannte Konstruktionen von Wavelet-Basen verwendet werden und die Vereinigung dieser Wavelet-Basen ergibt unter schwachen Voraussetzungen an die Gebietszerlegung einen Wavelet-Frame für das gesamte Gebiet, also ein im Allgemeinen redundantes, aber dennoch stabiles Erzeugendensystem. Dabei entfällt die Problematik, Wavelets an den Grenzen von Teilgebieten zusammenfügen zu müssen. Obwohl das so erzeugte Ansatzsystem redundant ist, konnten dennoch adaptive Wavelet-Verfahren für lineare Probleme konstruiert werden, die mit optimaler Rate konvergieren, siehe etwa [33, 34, 35, 104]. Als besonders leistungsfähig haben sich dabei die adaptiven Schwarz-Gebietszerlegungsverfahren aus [111, 123] erwiesen. Die grundlegende Idee solcher Verfahren ist es, das Problem auf dem Gebiet auf eine Folge von Teilproblemen auf den Teilgebieten zurückzuführen. Dieser Ansatz ist schon seit Längerem unabhängig von Wavelet-Verfahren bekannt, siehe beispielsweise [59, 60, 96, 103, 115, 126]. Die Kombination mit adaptiven Wavelet-Verfahren erwies sich aber als besonders effizient, unter anderem weil solche Verfahren auf natürliche Art mit der Konstruktion des aggregierten Wavelet-Frames mittels überlappender Gebietszerlegung harmonieren und damit auf den Teilgebieten Riesz-Basen vorhanden sind. Somit sind zumindest auf Ebene der Teilprobleme keine Redundanzen in der Darstellung der Lösung zu befürchten.

Die Anwendung von Wavelet-Frame-Verfahren ist darüber hinaus auch bei einer Klasse nichtlinearer elliptischer Problemen möglich. In [75] wurde ein auf einer Richardson-Iteration basiertes Verfahren eingeführt, das beweisbar konvergent und asymptotisch optimal ist. Im Hinblick auf die Verbesserungen, die schon bei linearen Problemen durch die Verwendung von Gebietszerlegungsverfahren vom Schwarz-Typ möglich waren, erscheint es jedoch vielversprechend, auch für nichtlineare Probleme adaptive Wavelet-Gebietszerlegungsverfahren zu untersuchen. Dies ist der Fokus der vorliegenden Arbeit, wobei Teile der Ergebnisse bereits in [38, 82] veröffentlicht wurden.

Ein zentrales Ziel dieser Dissertation ist also die Entwicklung und Analyse *adaptiver Wavelet-Gebietszerlegungsverfahren* für eine Klasse *nichtlinearer elliptischer Probleme*. Dazu werden wir zunächst, basierend auf einem Ansatz mit einer einfachen Linearisierungsstrategie aus [85], idealisierte Schwarz-Verfahren für nichtlineare Probleme einführen. Eine solche Linearisierung erscheint notwendig, um schnelle lokale Löser implementieren zu können. Dabei werden wir sowohl *multiplikative* wie auch *additive Schwarz-Verfahren* betrachten. Bei Ersteren sind die Teilprobleme sequentiell zu lösen, bei Letzteren können diese parallel bearbeitet werden. Aus den idealisierten Verfahren, welche in dieser Form noch nicht implementiert werden können, werden wir mithilfe bekannter Bausteine schrittweise adaptive Wavelet-Verfahren entwickeln. Wir beweisen, dass die so definierten Verfahren konvergent sind, sofern der nichtlineare Teil der Gleichung einer gewissen Kontraktionsbedingung genügt. Darüber hinaus werden wir nachweisen, dass die entwickelten Verfahren unter technischen Voraussetzungen asymptotisch optimal sind, also die Konvergenzrate der besten N-Term-Approximation erreichen. Zusätzlich zeigen wir, dass die Verfahren einer analogen Abschätzung für den benötigten Rechenaufwand genügen, also *lineare Komplexität* in der Anzahl der verwendeten Freiheitsgrade aufweisen. Da die theoretisch bewiesenen Ergebnisse asymptotischer Natur sind, ist es wichtig, anhand numerischer Experimente nachzuweisen, dass das erwartete Verhalten auch in der Praxis bei realistischen Größenordnungen beobachtbar ist. Daher werden in dieser Arbeit numerische Tests an repräsentativen Problemen in ein und zwei Raumdimensionen durchgeführt. Es zeigt sich, dass die erwartete Konvergenzrate tatsächlich realisiert wird und die Vorteile gegenüber gewöhnlichen uniformen Verfahren werden erkennbar.

Ein weiterer Aspekt dieser Arbeit ist es, die Anwendbarkeit der hier vorgestellten Prinzipien auf andere Gleichungen oder Problemklassen zu untersuchen. In [86] wurde ein idealisiertes additives Schwarz-Verfahren für ein wichtiges Beispiel aus der Strömungslehre, der inkompressiblen, stationären Navier-Stokes-Gleichung, entwickelt und dessen Konvergenz zumindest für kleine Reynolds-Zahlen gezeigt. Da die Konvergenztheorie auf einer schwachen Formulierung mit einem divergenzfreien Ansatzraum beruht, erscheint die Kombination dieses idealisierten Verfahrens mit divergenzfreien Wavelets, siehe [83, 109, 119], sinnvoll. Wir formulieren daher ein additives Wavelet-Gebietszerlegungsverfahren für dieses Problem, und zeigen dessen Konvergenz. Darüber hinaus können wir nachweisen, dass das so konstruierte Verfahren asymptotisch optimal in Bezug auf die Anzahl der Freiheitsgrade der Näherungen ist.

Die bisher genannten, in dieser Arbeit konstruierten Verfahren, nutzen eine einfache Linearisierungsstrategie, bei der die Nichtlinearität in jedem Schritt mit einer konstanten Funktion approximiert wird. Damit dies zu einem beweisbar konvergenten Algorithmus führt, sind jedoch recht strenge Annahmen an die Nichtlinearität zu stellen. Um eine größere Klasse nichtlinearer Probleme abzudecken, erscheint es daher sinnvoll, eine andere Linearisierungsstrategie zu betrachten, wie etwa im Newton-Verfahren. Ein adaptives Wavelet-Newton-Verfahren wurde in [27] vorgestellt, dieses verlangt jedoch eine Wavelet-Basis auf dem Gebiet, was zu oben diskutierten Schwierigkeiten führen kann. Um diese zu umgehen und den Ansatz des Newton-Verfahrens mit den Vorteilen eines überlappenden Gebietszerlegungsverfahrens zu kombinieren, entwickeln wir im letzten Teil dieser Arbeit ein adaptives Wavelet-Newton-Schwarz-Verfahren. Wir beweisen dessen Konvergenz für monotone Probleme und zeigen die asymptotische Optimalität in Bezug auf die Anzahl der Freiheitsgrade. Darüber hinaus belegen wir diese Ergebnisse mit numerischen Tests. Auch wenn in Bezug auf die Rechenzeit dieses Verfahren im Vergleich zu den oben dargestellten einfachen Schwarz-Verfahren aufwändiger ist, so scheint es ein größeres Potential zur Anwendung auf allgemeinere nichtlineare Probleme zu besitzen.

Abschließend zeigen wir in dieser Arbeit noch einige Perspektiven für zukünftige Weiterentwicklungen auf.

Contents

Introduction

Many problems in the natural sciences, such as in physics and biology, or in the social sciences, such as in economics, can be described with models that involve *boundary value problems* with *partial differential equations*. Moreover, realistic models often lead to nonlinear equations, of which linear models are only a rough approximation. For such equations, it is usually not possible or practical to find an explicit solution. Therefore, the numerical solution of partial differential equations is an important research area within the field of numerical analysis. The treatment of these problems depends strongly on the type of the equation, because different types may have very different characteristics. In this thesis, we will we concerned with a sort of *nonlinear elliptic equations*, which are typically used to describe stationary problems. An example is the nonlinearly perturbed Poisson equation

$$-\Delta u(x) + G(u(x)) = f(x),\ x \in \Omega, \quad u(x) = 0,\ x \in \partial\Omega,$$

where $G : \mathbb{R} \to \mathbb{R}$ is a given function such as $G(x) = x^3$. Using the weak formulation of boundary value problems, such equations can be written as

$$\mathcal{L}u + \mathcal{G}(u) = f, \tag{0.1}$$

where $\mathcal{L}$ is a linear elliptic operator between a Hilbert space $\mathcal{H}$ and its dual $\mathcal{H}'$, $\mathcal{G}$ is a nonlinear mapping between $\mathcal{H}$ and $\mathcal{H}'$ and f is a given right-hand side in $\mathcal{H}'$. In most parts of this thesis, $\mathcal{H}$ will be a Sobolev space on a bounded domain $\Omega \subset \mathbb{R}^d$ including boundary conditions, $\mathcal{H} = H_0^t(\Omega)$. A standard approach for the numerical treatment of these problems, both linear and nonlinear, is a Galerkin method with uniform finite elements. The basic idea of this kind of method is to project the problem onto a finite-dimensional subspace generated by finite element functions constructed on an equidistant grid on the domain. To improve the quality of the approximation, the grid is uniformly refined. For a detailed description of such methods, their application and background, we refer, e.g., to [62, 68, 80]. These methods are typically easy to implement and they are well-understood. However, they suffer from a few drawbacks. For standard finite element methods for the Poisson equation, the condition number of the finite-dimensional system grows quadratically with the inverse of the mesh size. This problem, however, can be tackled using suitable preconditioning strategies. A prominent example are *multigrid methods*, where, roughly speaking, an approximation on a coarse grid is used as a preconditioner for the calculation on a finer grid. For an overview of such multigrid methods, we refer, e.g., to [11, 69]. Another principal difficulty with uniform methods may occur when the solution of the equation is not smooth. This is often the case due to singularities, that can, for instance, be imposed by the right-hand side or by the geometry of the domain. Such singularities can significantly reduce the convergence rate of uniform methods, understood as the rate between the error and the number of degrees of freedom of the approximation.

To develop more efficient methods for problems with non-smooth solutions, *adaptive methods* have increasingly come into focus. The basic idea of these methods is to use a higher resolution in the parts of the domain where the solution has singularities. In the parts where the solution is smooth, a coarser resolution can be chosen. For finite element discretizations, locally refining the grid can also be combined with adapting the order of the elements, see, e.g., [102]. While some of these methods require a-priori information on the solution, in fully adaptive methods, the choice where the higher resolution is applied is made by the algorithm, so no a-priori information on the solution is needed. To achieve this, it is necessary to construct local error estimators. Compared to standard uniform methods, this leads to more complex methods. On the other hand, they often realize a higher convergence rate. For results in this direction, we refer, e.g., to [9, 91, 106]. The theoretical proof of such convergence rates, however, can be technical and is often restricted to limited settings. For a general overview about adaptive finite element methods, we also refer to [120].

A relatively new approach to the numerical solution of partial differential equations and an alternative to adaptive finite element methods is the construction of adaptive methods based on a *wavelet discretization*. With these methods, we will see that, in a very broad setting, convergence rates can be shown that are optimal in a sense we will explain below.

Adaptive Wavelet methods

Starting in the 1980s, wavelets have become a vital tool in numerical analysis. Originally, they were mostly applied in signal processing and time-frequency analysis. In recent years, the adaptive solution of operator equations using wavelets has also come into focus. For a general introduction to wavelets and an overview about early applications, we refer, e.g., to [20, 48, 90, 125]. The basic idea for the construction of wavelet bases on the real line is to find a *mother wavelet* $\psi \in L_2(\mathbb{R})$ so that, simply by dilating and translating this function, i.e., by setting

$$\Psi^{L_2(\mathbb{R})} := \{\psi_{j,k}(x) := 2^{j/2}\psi(2^j x - k), \, j, k \in \mathbb{Z}\},$$

we obtain an orthogonal basis, or at least a Riesz basis, for $L_2(\mathbb{R})$. If ψ is compactly supported, increasing the parameter j leads to functions with smaller support, that are particularly able to resolve details or local phenomena. In this sense, this parameter j is also called the *level* of $\psi_{j,k}$ and the representation of a function in a wavelet basis can also be seen as a multiscale representation. Wavelet bases can also be defined for bounded intervals or even bounded domains. In this case, it becomes necessary to adapt the wavelets around the boundary, so the structure becomes a little more complicated, but the basic idea remains as described above. Moreover, wavelet bases can be constructed with additional features that make them particularly attractive for the numerical treatment of partial differential equations, see, e.g., [14, 15, 44, 45, 95]. The wavelets designed in these works are compactly supported, they are piecewise polynomial and are, depending on the order of the polynomials, several times continuously differentiable on the whole domain. Moreover, they have vanishing moments. With these properties combined, it can be shown that these wavelet bases characterize a whole range of classical smoothness spaces, in particular Sobolev and Besov spaces. This means that the norm of a function in such spaces is equivalent to a weighted sequence norm of its the expansion coefficients of the function in the basis. In particular, the characterization of Sobolev spaces implies that by a simple rescaling, we have available wavelet Riesz bases $\Psi = \{\psi_\lambda\}_{\lambda \in \Lambda}$ for the Sobolev spaces that arise in the weak formulation of partial differential equations. This allows for the reformulation of the problem (0.1) as an equivalent problem

$$\mathbf{Au} + \mathbf{G}(\mathbf{u}) = \mathbf{f} \tag{0.2}$$

in the space $\ell_2(\Lambda)$ of square summable sequences. Here, the infinite-dimensional matrix $\mathbf{A}$ corresponds to the linear part $\mathcal{L}$, $\mathbf{G}$ is the counterpart of the nonlinearity $\mathcal{G}$ and $\mathbf{f}$ is the right-hand side in wavelet coordinates. With the properties of the wavelet bases, it can be shown that the matrix $\mathbf{A}$ and all its submatrices have uniformly bounded condition numbers and that the entries decay exponentially away from the main diagonal. This has the consequence that this matrix is *compressible*, which means that it can be well approximated by sparse matrices. This observation lies at the core of the design of an efficient approximate matrix-vector product, and, using this method, adaptive wavelet schemes for linear problems in the form $\mathcal{L}u = f$, or their discrete counterparts $\mathbf{Au} = \mathbf{f}$. The first adaptive wavelet Galerkin and Richardson methods based on these principles were introduced in [25, 26]. In these works, it could be verified that the algorithms are convergent and that they are *asymptotically optimal*. This means that they reproduce the convergence rate of the *best N-term approximation* of the exact solution $\mathbf{u}$ from (0.2), i.e., the best approximation using N degrees of freedom. In other words, the estimate

$$\inf\{\|\mathbf{u} - \mathbf{v}\|_{\ell_2(\Lambda)}, \ \mathbf{v} \in \ell_2(\Lambda), \ \#\operatorname{supp}\mathbf{v} \leq N\} \lesssim N^{-s} \tag{0.3}$$

implies that the algorithm also converges with rate s with respect to the number of degrees of freedom, at least for a certain range of parameters s. Here and throughout this thesis, $a \lesssim b$ means that $a \leq \bar{C} \cdot b$ holds with a constant $\bar{C} > 0$ independent of the parameters that a and b may depend on. Similarly, $a \gtrsim b$ is to be read as $b \lesssim a$, and $a \approx b$ means that both $a \lesssim b$ and $a \gtrsim b$ hold. In addition to the estimate of the degrees of freedom, it could be shown that the algorithms require a number of computational operations that is proportional to the degrees of freedom involved, hence they have *linear complexity*. These results are particularly important because the rate s in (0.3) can be characterized in terms of the smoothness of the solution

u in a scale of *Besov spaces*, see, e.g., [23, 52], and it has turned out that regularity measured in this *adaptivity scale* is ofter higher than the classical Sobolev regularity. In particular, this is the case if the solution has local singularities that typically occur on domains with corners, see [30, 32, 39]. Therefore, it is theoretically backed that the application of adaptive wavelet methods in many cases pays off in terms of a higher convergence rate.

More recently, starting from [27], the treatment of nonlinear problems of the form (0.1) with adaptive wavelet methods has come into focus. The general framework suggested there covers a broad range of methods including an inexact Richardson iteration as well as Newton's method. A main difficulty in the realization of these methods is the efficient evaluation of the nonlinearity $\mathbf{G}(\cdot)$ from (0.2), so that the linear complexity of the algorithm is achieved. This difficulty can be overcome when working with a *tree structure* on the set of wavelet indices, see, in addition to [27], also [7, 10, 24, 28]. With this structural modification, similar convergence and optimality results as for linear problems could be shown. Although the benchmark, other than in (0.3), is now the N-term approximation restricted to tree-structured vectors, the optimal approximation rate s can still be described in terms of the regularity of the solution u in almost the same scale of Besov spaces. Hence, the theoretical justification for the application of adaptive wavelet methods remains intact.

Adaptive wavelet methods using wavelet bases have been studied for very different kinds of other problems including integral equations, see [43, 70, 71] as a narrow choice, or saddle point problems, see, e.g., [31, 36]. Quite recently, the treatment of stochastic partial differential equations with adaptive wavelet methods has come into focus, see, e.g., [21, 78]. For a general overview over adaptive wavelet methods and their applications, we also refer to [41, 107, 119].

Wavelet Frame methods and Domain Decomposition

So far, we have given an overview of adaptive methods using wavelet *bases*. The application of such methods therefore relies on the availability of a suitable wavelet basis. Although there is a broad range of possible constructions, most of them are tailored to the d-dimensional unit cube. On more general domains, such as non-convex polygonal domains, the realization involves some difficulties. A typical construction is done by splitting the domain into nonoverlapping subdomains that are diffeomorphic to the unit cube and by assembling wavelet bases on these subdomains, where known constructions can be applied, see, e.g., [14, 15, 46]. However, these wavelet bases have to, in a way, be pieced together along the interfaces of the subdomains. This process is technically rather complicated and may lead to a lower global regularity of the resulting wavelets or to worse condition number of the bases.

An alternative approach suggested in [104] is to resort to an *overlapping domain decomposition*. In many cases, the domain Ω can be decomposed into such overlapping subdomains Ω_i, that are diffeomorphic to the unit cube. Then, wavelet bases with Dirichlet boundary conditions on the subdomains are constructed, without the difficulty of piecing together wavelets along the interfaces. Under relatively weak conditions on the decomposition, simply collecting the bases on the subdomains results in an *aggregated wavelet frame* for the Sobolev space on the entire domain. Roughly speaking, a frame is a generating system that is possibly redundant, but stable. For a general overview over frames, we refer to [18]. In the case of aggregated wavelet frames, the redundancies are caused by the overlap of the subdomains. Although such a wavelet frame is often easier to construct than a basis, when applying frames for the solution of partial differential equations, one has to make sure the redundancies do not endanger the efficiency of the algorithm. Moreover, Galerkin methods can no longer be directly applied, because a finite subset of the frame may be close to or even linearly dependent, which has the submatrices of the stiffness matrix $\mathbf{A}$ are ill-conditioned or even singular. However, with an iterative solver such as an inexact Richardson iteration, an adaptive wavelet frame method was constructed in [104]. Moreover, with a few additional measures to control the redundancies, this algorithm was shown to converge with an asymptotically optimal rate. Over the course of time, further methods and techniques for the solution of linear equations using wavelet frames have been developed and analyzed, see, for instance, [33, 34, 35, 97].

For such linear problems, a particularly convincing approach from a theoretical and computational point of view was suggested in [111, 123]. There, the discretization with a wavelet frame was combined with classical Schwarz domain decomposition methods. The basic principles of such methods have been developed independent from wavelets, starting more than a century ago as a theoretical tool to derive existence results for the solutions of partial differential equations, see [103]. In recent decades, these methods have increasingly been studied for the numerical solution of partial differential equations, see [59, 60, 84, 126] as a small sample of the broad literature in this area, [96, 115, 124] for overviews and [63] for a nice presentation of the historical development of Schwarz methods. The basic idea is to reduce the problem posed on the domain Ω into a sequence of subproblems on the subdomains Ω_i. There are two main variants of such Schwarz methods. In the *multiplicative Schwarz method*, the subproblems need to be solved sequentially, so that the information gained from the solution of one subproblem can be used when setting up the subproblem on the next subdomain. In the *additive Schwarz methods*, the subproblems are constructed in a way so that they are independent from each other and can therefore be solved in parallel. This is particularly attractive given the architecture of modern multi-core and multi-processor computers. Moreover, the design of adaptive wavelet methods based on Schwarz methods leads to particularly sparse and efficient approximations of the solution, compare the numerical results in [123].

Furthermore, adaptive wavelet frame methods can also be applied for the treatment of *nonlinear problems* of the form (0.1). To do so, the evaluation routines for nonlinear expressions from [7, 27, 28] and the concept of tree structures have to be adapted to aggregated wavelet frames. Quite recently, these tools and concepts and, based on them, the construction of an asymptotically optimal Richardson iteration for monotone nonlinear problems were studied in [75, 76]. Hence, adaptive wavelet frame methods are available for a whole range of both linear and nonlinear problems. As an example, see also [17] for an application to *nonlinear inequalities*. However, the treatment of nonlinear problems remains challenging from a computational point of view.

Main objectives

Because already in the linear case, the application of Schwarz methods lead to very convincing results, both in theory and practice, it seems promising to study the construction of *adaptive wavelet Schwarz methods for nonlinear problems*. The theory of Schwarz methods for nonlinear problems and the application within finite element methods has been considered in a whole range of works, see, e.g., [12, 58, 72, 85, 87, 92] for methods and convergence results in different settings. In particular, we want to mention the analysis provided in [85], where it has been observed that a simple linearization strategy leads to a convergent algorithm given additional assumptions on the nonlinear term. However, to the best of our knowledge, the adaption of these ideas to the framework of adaptive wavelet methods, the convergence and optimality analysis of the adaptive algorithms as well as their implementation have not yet been explored. This is the main focus of this thesis and the papers [38, 82], which include parts of the results presented here. The tasks we are concerned with in this thesis can be summarized as follows.

(T1) A central objective is the development of adaptive wavelet Schwarz methods for nonlinear problems of the form (0.1). For this task, we want to adapt the basic strategy from [85] to the wavelet case. We shall show that the newly designed method is convergent and asymptotically optimal with respect to the degrees of freedom and the computational effort.

This task covers already the main *theoretical* objectives we want to achieve. However, the optimality results are of an asymptotic nature and the constants involved in these estimates can usually not be explicitly quantified. Therefore, we have to verify that the asymptotic behavior can actually be observed in practice.

(T2) We want to provide numerical experiments with representative test problems to show that the methods can be practically realized and that the expected convergence rates can be observed at realistic scales.

It is known from [86] that very similar principles as applied to the semilinear problems can be used to formulate an idealized additive Schwarz method for an important problem in fluid dynamics, the stationary, incompressible Navier-Stokes equation, at least if the Reynolds number is sufficiently small. By applying divergence-free wavelets, we want to adapt the principles applied for (T1) to this problem.

(T3) Another objective is develop an adaptive wavelet Schwarz method for the stationary Navier-Stokes equation. We want to prove its convergence and asymptotic optimality.

Furthermore, we are interested in whether adaptive Schwarz methods can be coupled with other strategies for the solution of nonlinear problems. A natural further step is to study the combination with Newton's method, compare [27]. The inexact Newton method presented there allows between linear and quadratic error reduction in each step. Although Newton-Schwarz methods have been studied before, to the best of our knowledge, an adaptive Newton-Schwarz method using wavelet frames has not yet been constructed. In the future, methods of this kind might show a larger potential for the generalization to more general nonlinear problems than (0.1), because Newton's method itself is known to converge, at least locally, for a broader range of equations.

(T4) Another objective is the formulation of an implementable adaptive wavelet Newton-Schwarz method, where the linear systems arising in each step of Newton's method are solved approximately with an additive Schwarz method. We want to show that even this method is convergent and asymptotically optimal. Moreover, we shall provide numerical experiments to observe the theoretically predicted results in practice.

In the following, we outline how these objectives will be addressed in this thesis.

Overview and layout of the thesis

In view of the above tasks, let us now give an outline over the contents provided in this thesis. As a rough guideline, the first three chapters lay the foundation for the construction of our methods by presenting and explaining mostly known results needed for the later analysis. In the following chapters, in a stepwise approach, we address tasks (T1) to (T4) by constructing, analyzing and testing adaptive wavelet methods fulfilling the requirements specified previously. In detail, the thesis is structured as follows.

In **Chapter 1**, we introduce the scope of problems we will work with. In particular, we introduce some important classes of domains and Sobolev spaces defined on such domains. We discuss their main properties and apply them to present a weak formulation of semilinear boundary value problems. In the course of this, we recall some important results on nonlinear mappings between Sobolev spaces. Afterwards, we give some examples of problems that can be treated within this framework. Finally, we discuss the connection to the stationary Navier-Stokes equation.

Afterwards, in **Chapter 2**, we review the construction of aggregated wavelet frames and explain how they are applied for the discretization of operator equations, i.e., for the reformulation of the problems as equations in the form (0.2) in the sequence space $\ell_2(\Lambda)$. To do so, in Section 2.1 we introduce Riesz bases and frames in Hilbert spaces.

In particular, we explain the concept of *Gelfand frames*, that can be used to describe frames that are tailored simultaneously to Lebesgue spaces, in which the wavelets are originally constructed, and the Sobolev spaces arising in the weak formulation. After the introduction of these rather abstract concepts, we outline in Section 2.2 known construction principles for the design of wavelet bases and aggregated wavelet frames. Starting from the situation on real line, we sketch the design principles of wavelet bases via the important concept of a *multiresolution analysis*. Then, we describe the modifications needed on bounded intervals. A practical construction of spline wavelets along these lines is sketched afterwards. The constructions on the interval can, by a tensor product approach, be expanded to the unit cube. By means of an overlapping domain decomposition, with the wavelet bases on the unit cube, we outline the construction of an aggregated wavelet frame. Moreover, we explain the design of divergence-free wavelets that are suitable for the treatment of the incompressible Navier-Stokes equation. In Section 2.3, we describe how the aggregated wavelet frames can be applied for the discretization of the problem (0.1) or the Navier-Stokes equation.

Chapter 3 is concerned with the approximation of functions using wavelets and the connection between approximation rates and the regularity of the function to approximate. We start with giving a rough introduction to the basics of approximation theory and to Besov spaces. We thereafter explain in Section 3.2 how the convergence rates of linear and nonlinear approximation schemes with wavelet bases can be described in terms of Sobolev and Besov regularity, respectively. Afterwards, these ideas are connected with the discretized equation (0.2) and the concept of best N-term approximation. Since nonlinear problems make it almost unavoidable to work with tree-structured index sets, we introduce such structures and discuss the properties of the best N-term approximation when restricted to trees. As we will explain, this optimal approximation serves as a benchmark for our adaptive algorithms and allows us to introduce the concept of asymptotic optimality. Finally, we collect regularity results for the solutions of linear and nonlinear partial differential equations. These results illustrate the potential benefits when applying asymptotically optimal algorithms for numerical solution of these problems.

The first main objective (T1) of this thesis is addressed in **Chapter 4**. We proceed stepwise and start with constructing idealized multiplicative and additive Schwarz algorithms. Thereafter, we show their convergence, given that the nonlinearity fulfills a contraction property. We furthermore prove that the convergence results hold true if the subproblems are only solved inexactly. To formulate these algorithms as adaptive wavelet methods, in Section 4.2 we explain and collect all necessary building blocks needed to evaluate the terms arising in the discretized form (0.2) of our problem. In addition, we discuss a coarsening method that will help us keep an optimal balance between the size of the iterates in terms of the degrees of freedom and the accuracy of the approximation. In Section 4.3, these building blocks will be applied to construct an adaptive wavelet multiplicative Schwarz method. We discuss in detail how to solve the subproblems and show that the method is convergent and asymptotically optimal.

This is a central result of this thesis, summarized in Theorem 4.3.10. Analogously, we provide an adaptive wavelet version of the *additive* Schwarz method in Section 4.4. In Theorem 4.4.3, we show convergence and asymptotic optimality of this version.

In **Chapter 5**, we address Task (T2) and provide numerical experiments to demonstrate that the convergence and asymptotic behavior of the algorithms can be practically observed. We first explain the realization of the methods. Thereafter, we present detailed numerical experiments in one in two space dimensions with representative examples, in which all important effects can be observed. We conclude this chapter with some remarks and tests to show how the methods can further be accelerated.

Chapter 6 is devoted to Task (T3), the design of an adaptive additive Schwarz method for the stationary, incompressible Navier-Stokes equation using a divergence-free wavelet frame. We start with presenting an idealized version of the algorithm and give a proof of its convergence. Afterwards, we discuss a strategy to avoid redundancies in the overlapping regions of the subdomains, that is compatible with the application of divergence-free wavelets. With this strategy at hand, we are ready to formulate an adaptive wavelet version of the method and show that it is convergent and asymptotically optimal with respect to the degrees of freedom. Furthermore, we outline how to solve the subproblems on the subdomains arising in the algorithm.

The construction of an adaptive wavelet Newton-Schwarz method, summarized in Task (T4), is addressed in **Chapter 7**. We proceed as with the other methods and first describe the basic principles of the algorithm. Thereafter, we formulate an adaptive wavelet version of the method, which allows between linear and quadratic error reduction in each step. We prove that even this method is convergent and asymptotically optimal with respect to the degrees of freedom. Moreover, we explain the numerical solution of the subproblems. The results are supported by representative numerical experiments on the L-shaped domain in two space dimensions.

Chapter 1

Semilinear Elliptic Equations

In this thesis, we will be concerned with the numerical treatment of boundary value problems with semilinear partial differential equations. A typical example is the nonlinearly perturbed Poisson equation

$$\Delta u + u^3 = f$$

with zero boundary conditions. We will see in Sections 1.2 and 1.3, that, with some preparations, many semilinear partial differential equations can be treated in a common setting as operator equations between a Hilbert space and its dual. Therefore, we first introduce this rather abstract setting and then show how the partial differential equations can be embedded into this setting. This has the advantage that it allows us to use techniques from functional analysis and the theory of function spaces to apply existence, uniqueness and regularity results. Moreover, this setting might be suitable even for other kinds of equations such as integral equations, which, however, are not the main focus of this thesis. The semilinear operator equations we will work with are of the form

$$\mathcal{L}u + \mathcal{G}(u) = f, \tag{1.1}$$

where $\mathcal{L}$ is a linear operator from a Hilbert space $\mathcal{H}$ into its dual $\mathcal{H}'$, that is induced by a symmetric bilinear form

$$(\mathcal{L}u)(v) := a(u, v) = a(v, u), \quad u, v \in \mathcal{H}.$$

We assume that $a(\cdot, \cdot)$ is *bounded*, $|a(u, v)| \lesssim \|u\|_{\mathcal{H}} \cdot \|v\|_{\mathcal{H}}$, and *elliptic*, $a(v, v) \gtrsim \|v\|_{\mathcal{H}}^2$. Under these conditions, the induced operator $\mathcal{L}$ is elliptic as well, which means that

$$\|\mathcal{L}v\|_{\mathcal{H}'} \approx \|v\|_{\mathcal{H}}. \tag{1.2}$$

Furthermore, we assume that the second part $\mathcal{G}$ is a nonlinear mapping from $\mathcal{H}$ to $\mathcal{H}'$ and f is a given right-hand side in $\mathcal{H}'$. The bilinear form $a(\cdot, \cdot)$ allows us to define an *energy norm* on $\mathcal{H}$ by $\|v\|_{\mathcal{H}} := a(v, v)^{1/2}$, which is equivalent to the standard norm on $\mathcal{H}$. The induced operator norm on $\mathcal{H}'$ will be written as

$$\|f\|_{\mathcal{H}'} := \sup_{v \in \mathcal{H}, v \neq 0} \frac{|\langle f, v \rangle_{\mathcal{H}' \times \mathcal{H}}|}{\|v\|_{\mathcal{H}}}.$$

Here and in the following, $\langle f, v \rangle_{\mathcal{H}' \times \mathcal{H}} := \langle v, f \rangle_{\mathcal{H} \times \mathcal{H}'} := f(v)$ denotes the *dual pairing* between $\mathcal{H}$ and $\mathcal{H}'$. If it is clear from the context which spaces are meant, we often omit the subscripts and simply write $\langle \cdot, \cdot \rangle$ or $\| \cdot \|$, respectively.

Because of the Cauchy-Schwarz inequality, it is $\| \mathcal{L} v \|_{\mathcal{H}'} = \| v \|_{\mathcal{H}}$. It is known that for linear problems, $\mathcal{G} \equiv 0$, the above conditions are sufficient to guarantee that the operator $\mathcal{L}$ is boundedly invertible, i.e., that the problem $\mathcal{L} u = f$ has a unique solution u which depends continuously on f, see [68, Th. 6.5.9]. To show existence and uniqueness for nonlinear problems such as (1.1), however, additional assumptions are required. We consider two types of nonlinearities for which such results hold, namely nonlinearities that are contractive, and monotone nonlinearities with at most polynomial growth.

The first and easier type can be treated right away in the abstract operator setting. The discussion of the second type will be postponed to Section 1.2, because it is closer related to the concrete Sobolev spaces appearing in the weak formulation of elliptic partial differential equations.

Lemma 1.0.1. *Assume there exists a constant $c < 1$ such that*

$$\| \mathcal{G}(v) - \mathcal{G}(w) \|_{\mathcal{H}'} \leq c \| v - w \|_{\mathcal{H}} \tag{1.3}$$

for all $v, w \in \mathcal{H}$. Then, Equation (1.1) has a unique solution $u \in \mathcal{H}$. Moreover, the solution depends continuously on the right-hand side, i.e., for solutions u and $\tilde{u}$ corresponding to the right-hand sides f and $\tilde{f}$, respectively, it is

$$\| u - \tilde{u} \|_{\mathcal{H}} \leq \frac{1}{1 - c} \| f - \tilde{f} \|_{\mathcal{H}'} \tag{1.4}$$

Proof. As in [85], we see that the mapping

$$\phi : \mathcal{H} \to \mathcal{H}, \quad v \mapsto \mathcal{L}^{-1}(f - \mathcal{G}(v))$$

is a contraction because of

$$\| \phi(v) - \phi(w) \|_{\mathcal{H}} = \| \mathcal{L}^{-1}(\mathcal{G}(w) - \mathcal{G}(v)) \|_{\mathcal{H}} = \| \mathcal{G}(v) - \mathcal{G}(w) \|_{\mathcal{H}'} \leq c \| v - w \|_{\mathcal{H}}.$$

Hence, by the Banach fixed point theorem, there exists a unique fixed point $u \in \mathcal{H}$, which, in turn, is the unique solution to (1.1).

Let $f, \tilde{f} \in \mathcal{H}'$ be two right-hand side functions. Then, for the corresponding solutions u and $\tilde{u}$ to (1.1), it holds that

$$u - \tilde{u} + \mathcal{L}^{-1}(\mathcal{G}(u) - \mathcal{G}(\tilde{u})) = \mathcal{L}^{-1}(f - \tilde{f}).$$

Hence, by the reverse triangle inequality, we obtain

$$\| f - \tilde{f} \|_{\mathcal{H}'} = \| \mathcal{L}^{-1}(f - \tilde{f}) \|_{\mathcal{H}} \geq \| u - \tilde{u} \|_{\mathcal{H}} - \| \mathcal{L}^{-1}(\mathcal{G}(u) - \mathcal{G}(\tilde{u})) \|_{\mathcal{H}} \geq (1 - c) \| u - \tilde{u} \|_{\mathcal{H}}.$$

This shows the estimate (1.4) and thereby completes the proof. $\qquad \square$

Remark 1.0.2. *Note that one application of the mapping ϕ amounts to solving a linear problem. By the Banach fixed point theorem, for any $u^{(0)} \in \mathcal{H}$, the iterates $u^{(n+1)} := \phi(u^{(n)})$ converge towards the solution u. Therefore, Lemma 1.0.1 is a starting point for some of the iterative schemes presented later.*

1.1 Domains and function spaces

To formulate an elliptic partial differential equation as an operator equation in the form (1.1), we first have to define the right function spaces to work with, namely *Sobolev spaces*. With the help of these function spaces, the weak formulation of a partial differential equation is introduced. A more detailed outline of Sobolev spaces can be found in [2, 116].

Let $U \subset \mathbb{R}^d$. We recall that a function $f : U \to \mathbb{R}^d$ is called *Lipschitz continuous*, or simply *Lipschitz*, if there exists a constant $K < \infty$ such that for all $x, y \in U$ it holds that

$$|f(x) - f(y)| \leq K|x - y|.$$

Definition 1.1.1. *A subset $\Omega \subset \mathbb{R}^d$ is called a* domain *if it is open and connected. A bounded domain Ω with boundary $\partial\Omega$ is called* Lipschitz domain *(C^k-domain) if for every $x \in \partial\Omega$ there exists a neighborhood U of x and a bijective mapping $\phi_x : U \to B_1(0) := \{z \in \mathbb{R}^d, |z| < 1\}$ such that ϕ_x and ϕ_x^{-1} are Lipschitz continuous (or in C^k, respectively) and*

$$\phi_x(U \cap \Omega) = \{z \in B_1(0), z_d > 0\},$$
$$\phi_x(U \cap \partial\Omega) = \{z \in B_1(0), z_d = 0\},$$
$$\phi_x(U \setminus \overline{\Omega}) = \{z \in B_1(0), z_d < 0\}.$$

A subset of a domain, that is a (Lipschitz) domain itself, is called (Lipschitz) subdomain.

Let Ω be a domain in $\mathbb{R}^d$. The *Lebesgue spaces* $L_p(\Omega)$ for $0 < p \leq \infty$ are defined as the spaces of all measurable functions $f : \Omega \to \mathbb{C}$, for which

$$\|f\|_{L_p(\Omega)} := \begin{cases} \left(\int\limits_\Omega |f(x)|^p \, \mathrm{d}x \right)^{1/p} & , \quad p < \infty, \\ \operatorname{ess\,sup}_{x \in \Omega} |f(x)| & , \quad p = \infty, \end{cases}$$

is finite. We consider functions in L_p as equal if they take the same value except possibly on a null set. With this modification, the spaces $(L_p(\Omega), \|\cdot\|_{L_p(\Omega)})$ are Banach spaces for all $1 \leq p \leq \infty$. The space $L_2(\Omega)$ is even a Hilbert space equipped with the inner product

$$\langle f, g \rangle_{L_2(\Omega)} := \int\limits_\Omega f(x)\overline{g(x)} \, \mathrm{d}x.$$

In the case $0 < p < 1$, the L_p-spaces are only quasi-Banach spaces, which are Banach spaces except that the triangle inequality is only true up to a constant

$$\|f + g\|_{L_p(\Omega)} \lesssim \|f\|_{L_p(\Omega)} + \|g\|_{L_p(\Omega)}.$$

For parameters $1 < p, q < \infty$, with $\frac{1}{p} + \frac{1}{q} = 1$, the well-known Hölder inequality states that if $f \in L_p(\Omega)$ and $g \in L_q(\Omega)$, it is $fg \in L_1(\Omega)$ with

$$\|fg\|_{L_1(\Omega)} \leq \|f\|_{L_p(\Omega)} \|g\|_{L_q(\Omega)}.$$

This inequality can be generalized to other parameters in the sense that if $f_i \in L_{q_i}(\Omega)$, $i = 1, \ldots, k$, and $1 \leq q_i \leq \infty$ with $\frac{1}{p} = \sum_{i=1}^{k} \frac{1}{q_i}$, then $f := \prod_{i=1}^{k} f_i \in L_p(\Omega)$ with

$$\|f\|_{L_p(\Omega)} \leq \prod_{i=1}^{k} \|f_i\|_{L_{q_i}(\Omega)}, \tag{1.5}$$

see, e.g., [2, Cor. 2.6]. Moreover, if $1 \leq p \leq q \leq \infty$ and Ω has finite measure, we have $L_q(\Omega) \subset L_p(\Omega)$ in the sense of a continuous embedding. The dual space of $L_p(\Omega)$, $1 < p < \infty$ is again a Lebesgue space, namely $L_q(\Omega)$ with $\frac{1}{p} + \frac{1}{q} = 1$, and, in addition, $L_1(\Omega)' = L_\infty(\Omega)$. The space $L_\infty(\Omega)'$, however, does not coincide with $L_1(\Omega)$. Hence, the spaces $L_1(\Omega)$ and $L_\infty(\Omega)$ are exceptional Lebesgue spaces in the sense that they are not reflexive. This is also why they often have to be treated separately.

We say that $U \subset\subset \Omega$, if the closure $\overline{U}$ of U is compact and a subset of Ω. Then, we can define the set $L_{p,loc}(\Omega)$ of functions that are contained in $L_p(U)$ for all $U \subset\subset \Omega$. Let $f \in L_{1,loc}(\Omega)$ and denote by

$$C_0^\infty(\Omega) := \{f \in C^\infty(\Omega), \operatorname{supp} f \subset\subset \Omega\}$$

the set of all *test functions*. Then, a function $f_\alpha \in L_{1,loc}(\Omega)$ is called *weak derivative* of f of order $\alpha \in \mathbb{N}_0^d$ if for all $\phi \in C_0^\infty(\Omega)$ it holds that

$$\int_\Omega f(x) D^\alpha \phi(x) \, \mathrm{d}x = (-1)^{|\alpha|} \int_\Omega f_\alpha(x) \phi(x) \, \mathrm{d}x \tag{1.6}$$

with $|\alpha| := \alpha_1 + \ldots + \alpha_d$, see [2, Section 1.60]. In this case, we write $D^\alpha f := f_\alpha$. It can be shown that the weak derivative is unique. Note that for a classical derivatives, Equation (1.6) holds by partial integration. Hence, a classical derivative is also a weak derivative, whereas the converse in not true in general.

Now, for $1 \leq p \leq \infty$ and $k \in \mathbb{N}$, we can define the *Sobolev spaces*

$$W^k(L_p(\Omega)) := \{f \in L_p(\Omega), D^\alpha f \in L_p(\Omega), |\alpha| \leq k\}$$

of all functions in $L_p(\Omega)$ with weak derivatives up to order k in $L_p(\Omega)$. We equip these spaces with the norms

$$\|f\|_{W^k(L_p(\Omega))} := \begin{cases} \left(\sum_{|\alpha| \leq k} \|D^\alpha f\|_{L_p(\Omega)}^p\right)^{1/p} & , \quad p < \infty, \\ \max_{|\alpha| \leq k} \|D^\alpha f\|_{L_\infty(\Omega)} & , \quad p = \infty. \end{cases}$$

By definition, it is $L_p(\Omega) = W^0(L_p(\Omega))$. The Sobolev spaces are Banach spaces, see [2, Th. 3.3]. For the special case $p = 2$, we even obtain the Hilbert space

$$H^k(\Omega) := W^k(L_2(\Omega))$$

with the scalar product

$$\langle f, g \rangle_{H^k(\Omega)} := \sum_{|\alpha| \leq k} \langle D^\alpha f, D^\alpha g \rangle_{L_2(\Omega)}.$$

An important property of Sobolev spaces is that spaces with higher Sobolev regularity k are embedded into spaces with lower regularity, but stronger $L_p(\Omega)$-norm. This is known as the Sobolev embedding, see, e.g., [2, Th. 4.12].

Theorem 1.1.2. *Let $\Omega \subset \mathbb{R}^d$ be a bounded Lipschitz domain, $k_1 > k_2 > 0$, $1 < p_1 \leq p_2 < \infty$ and $(k_1 - k_2)p_1 < d$. Then, it holds that*

$$W^{k_1}(L_{p_1}(\Omega)) \subset W^{k_2}(L_{p_2}(\Omega))$$

for all $k_1 - k_2 \geq d(\frac{1}{p_1} - \frac{1}{p_2})$.

In particular, on such bounded Lipschitz domains, we have the embedding

$$W^k(L_p(\Omega)) \subset L_{\frac{dp}{d-kp}}(\Omega) \tag{1.7}$$

for $d > kp$. If $d \leq kp$ we even have $W^k(L_p(\Omega)) \subset L_q(\Omega)$ for all $1 \leq q < \infty$, see, e.g., [49, Th. 2.72].

So far, we have defined Sobolev spaces of order $k \in \mathbb{N}$. However, it will sometimes be necessary to consider Sobolev spaces of fractional order $s \in \mathbb{R}^+ \setminus \mathbb{N}$. Therefore, for $s = k + \sigma$ with $k \in \mathbb{N}$ and $0 < \sigma < 1$, we define $W^s(L_p(\Omega))$, $p < \infty$ as the set of all functions in $L_p(\Omega)$, for which the norm

$$\|f\|_{W^s(L_p(\Omega))} := \left(\|f\|_{W^k(L_p(\Omega))}^p + \sum_{|\alpha|=k} \int_\Omega \int_\Omega \frac{|D^\alpha f(x) - D^\alpha f(y)|^p}{|x - y|^{d+p\sigma}} \, \mathrm{d}x \, \mathrm{d}y \right)^{1/p}$$

is finite. If $p \geq 1$, these spaces are again Banach spaces and $H^s(\Omega) := W^s(L_2(\Omega))$ is even a Hilbert space with scalar product

$$\langle f, g \rangle_{H^s(\Omega)} := \langle f, g \rangle_{H^k(\Omega)} + \sum_{|\alpha|=k} \int_\Omega \int_\Omega \frac{(D^\alpha f(x) - D^\alpha f(y))\overline{(D^\alpha f(x) - D^\alpha f(y))}}{|x - y|^{d+2\sigma}} \, \mathrm{d}x \, \mathrm{d}y.$$

Since would like to use Sobolev spaces for the formulation of partial differential equation, which generally require boundary conditions to determine a unique solution, we have to discuss how to incorporate them in the definition of suitable subspaces of

Sobolev spaces. Typically, we will be concerned with zero boundary conditions. It is not directly clear how to do define such boundary conditions, because the direct evaluation of a function $v \in W^k(L_p(\Omega))$ on a null set such as $\partial\Omega$ is not well-defined. One way to circumvent this difficulty and to define a subspace of a Sobolev space including boundary conditions is to introduce $W_0^s(L_p(\Omega))$ as the closure of all test functions in $W^s(L_p(\Omega))$ and set $H_0^s(\Omega) := W_0^s(L_2(\Omega))$. Alternatively, on Lipschitz domains and for $k \in \mathbb{N}$, we can make use of continuous *trace operators*

$$\mathrm{Tr} : W^k(L_p(\Omega)) \to W^{k-\frac{1}{p}}(L_p(\Omega)), \quad 1 < p < \infty,$$

that generalize the idea of the evaluation of functions on lower-dimensional subsets to Sobolev functions. As expected from a trace operator, it holds that $\mathrm{Tr}(v) = v_{|\partial\Omega}$ for $v \in C^\infty(\overline{\Omega})$. With the help of the these operators, the spaces $W_0^k(L_p(\Omega))$ on smooth or polygonal domains can be characterized as the subspaces of all $u \in W^k(L_p(\Omega))$ such that $\mathrm{Tr}(D^\alpha u) = 0$ for all $|\alpha| \leq k - 1$, see [67, Section 1.5].

In the weak formulation of boundary value problems, the dual spaces of the Sobolev spaces with boundary conditions will also play a central role. Hence, the dual of $H_0^s(\Omega)$, $s > 0$ is written as $H^{-s}(\Omega) := (H_0^s(\Omega))'$. Because $H_0^s(\Omega)$ is continuously and densely embedded in $L_2(\Omega) = (L_2(\Omega))'$, which in turn is continuously and densely embedded in $H^{-s}(\Omega)$, we obtain the embeddings

$$H_0^s(\Omega) \subset L_2(\Omega) \subset H^{-s}(\Omega). \tag{1.8}$$

In particular, the dual pair $\langle \cdot, \cdot \rangle_{H^{-s}(\Omega) \times H_0^s(\Omega)}$ is the continuous extension of the inner product on $L_2(\Omega)$. Such a construction $\mathcal{B} \subset \mathcal{H} \subset \mathcal{B}'$ with a Banach space $\mathcal{B}$, that is continuously and densely embedded in a Hilbert space $\mathcal{H}$, is generally called a *Gelfand triple*, written as $(\mathcal{B}, \mathcal{H}, \mathcal{B}')$. The Gelfand triple (1.8) will often be used implicitly when dealing with the weak formulation of elliptic equations and the construction of wavelet bases of frames for the discretization of such equations.

1.2 Weak formulation

In this section, we summarize the terminology needed to write a semilinear elliptic PDE as an operator equation between a Sobolev space and its dual and give further criteria for such an equation to have a unique solution. Throughout this section, let Ω be a Lipschitz domain. A differential operator of order $l = 2t$, $t \in \mathbb{N}$ is given by

$$L : C^{2t}(\Omega) \to C(\Omega), \; u \mapsto Lu := \sum_{|\alpha|,|\beta| \leq t} (-1)^{|\alpha|} D^\alpha (a_{\alpha,\beta} D^\beta u), \tag{1.9}$$

where $a_{\alpha,\beta}$ are coefficient functions that, for the time being, are assumed to be sufficiently smooth and symmetric, $a_{\alpha,\beta} = a_{\beta,\alpha}$. The operator is called *uniformly elliptic* if for all $x \in \Omega$ and $\xi \in \mathbb{R}^d$ it holds that

$$\sum_{|\alpha|=|\beta|=t} a_{\alpha,\beta}(x)\xi^{\alpha+\beta} \gtrsim |\xi|^{2t}.$$

The operator L will represent the linear part of the equation. To introduce a nonlinearity, let a function $G : \mathbb{R} \to \mathbb{R}$ be given. For any real-valued $v : \Omega \to \mathbb{R}$, we define the function $\mathcal{G}(v)$ simply by composition with G:

$$\mathcal{G}(v) : \Omega \to \mathbb{R}, \, v \mapsto G(v(x)).$$

The operator $\mathcal{G}$ that in this way maps a function v to a new function $\mathcal{G}(v)$ is called *Nemitsky operator*. We will later give criteria under which this is an operator between suitable function spaces. For further reading on Nemitsky operators, we also refer to [99].

Let, for the time being, $f \in C(\Omega)$. Then, with the notation defined above, the equation

$$Lu(x) + G(u(x)) = f(x), \quad x \in \Omega \tag{1.10}$$

will be called a *semilinear elliptic equation*, although the range of nonlinearities considered here is slightly restricted compared to the standard definition, where G may sometimes also depend on lower-order derivatives of u. In addition to Equation (1.10), we pose *Dirichlet boundary conditions* on the normal derivatives

$$\frac{\partial^k u}{\partial n^k} = \phi_k \text{ on } \partial\Omega, \quad k = 0, \ldots, t - 1, \tag{1.11}$$

where ϕ_k are given, continuous functions on $\partial\Omega$. In our case, we will work with *homogeneous boundary conditions*, namely $\phi_k \equiv 0$ for all $k = 0, \ldots, t - 1$. The two above equations together compose a *Dirichlet boundary value problem*. A function $u \in C^{2t}(\Omega) \cap C^{t-1}(\overline{\Omega})$ is called *classical* or *strong solution* of the boundary value problem if it solves the equations (1.10) and (1.11).

Unfortunately, the regularity assumptions on a strong solution are often too restrictive to guarantee its existence and uniqueness. This is especially the case if the right-hand side f or the domain Ω are not smooth. Therefore, it is advisable to resort to the weak formulation, which we will introduce now. The derivation of the weak formulation of linear problems is described more extensively in [68], whereas the weak formulation of semilinear elliptic equations and its link to minimization problems is outlined in [5].

To obtain the weak formulation, let us first integrate Lu against a test function $v \in C_0^\infty(\Omega)$. By partial integration, we obtain

$$\int_\Omega Lu(x) \cdot v(x) \, \mathrm{d}x$$

$$= \int_\Omega \sum_{|\alpha|,|\beta| \leq t} (-1)^{|\alpha|} D^\alpha (a_{\alpha,\beta}(x) D^\beta u(x)) v(x) \, \mathrm{d}x$$

$$= \sum_{|\alpha|,|\beta| \leq t} \int_\Omega a_{\alpha,\beta}(x) D^\alpha u(x) D^\beta v(x) \, \mathrm{d}x =: a(u, v). \tag{1.12}$$

17

Note that for the last expression to be well-defined, it is sufficient if u and v are contained in $H_0^t(\Omega)$. Therefore, the so-defined $a(\cdot, \cdot)$ is a symmetric bilinear form on $H_0^t(\Omega) \times H_0^t(\Omega)$. Moreover, although it was necessary to assume that the coefficients $a_{\alpha,\beta}$ are sufficiently smooth to perform partial integration and arrive at (1.12), when considering this equation only, we can relax this assumption. For the bilinear form to fit in our general setting, we need to make sure that $a(\cdot, \cdot)$ is bounded and elliptic. This is ensured by the following theorem.

Theorem 1.2.1. *Assume that $a_{\alpha,\beta} \in L_\infty(\Omega)$ for all $|\alpha|, |\beta| \leq t$, $a_{\alpha,\beta}$ is constant for $|\alpha| = |\beta| = t$, $a_{\alpha,\beta} = 0$ for $0 < |\alpha| + |\beta| \leq 2t - 1$ and $a_{0,0} \geq 0$. Let L be uniformly elliptic. Then $a(\cdot, \cdot)$ is bounded and elliptic on $H_0^t(\Omega)$.*

Proof. See [68, Th. 7.2.2 and Th. 7.2.7]. □

If these conditions are fulfilled, the operator

$$\mathcal{L} : H_0^t(\Omega) \to H^{-t}(\Omega), \ u \mapsto \mathcal{L}u := a(u, \cdot)$$

fits into the general setting described previously. To make sure that the nonlinear part $\mathcal{G}$ maps $\mathcal{H}$ into $\mathcal{H}'$ and that the overall problem is well-posed, a few more assumptions and results from [27] are required. First of all, let us remind that for a mapping $\mathcal{F}$ between a Hilbert space $\mathcal{H}$ and its dual $\mathcal{H}'$, the mapping $D\mathcal{F}(z) : \mathcal{H} \to \mathcal{H}'$ is called the *Fréchet derivative* of $\mathcal{F}$ in $z \in \mathcal{H}$, if

$$\langle v, D\mathcal{F}(z)w \rangle_{\mathcal{H} \times \mathcal{H}'} = \lim_{h \to 0} \frac{1}{h} \langle v, \mathcal{F}(z + hw) - \mathcal{F}(z) \rangle_{\mathcal{H} \times \mathcal{H}'} \tag{1.13}$$

holds for all $v, w \in \mathcal{H}$. We say that the Fréchet derivative is continuous if $D\mathcal{F}(v) \in \mathcal{H}'$ depends continuously on $v \in \mathcal{H}$.

Theorem 1.2.2. *Assume that G is* monotone,

$$(G(x) - G(y))(x - y) \geq 0, \quad x, y \in \mathbb{R},$$

and has at most polynomial growth, which means that for some $n^* \geq 1$ and $p \geq 0$, it is

$$|G^{(n)}(x)| \lesssim (1 + |x|)^{\max\{p-n,0\}}, \ n = 0, \ldots, n^*. \tag{1.14}$$

Then, if $t \geq \frac{d}{2}$ or $1 \leq p < \frac{d+2t}{d-2t}$, $\mathcal{G}$ maps $H_0^t(\Omega)$ into $H^{-t}(\Omega)$ and is stable, i.e., there exists a nondecreasing function $C_G : \mathbb{R} \to \mathbb{R}$ such that

$$\|\mathcal{G}(v) - \mathcal{G}(w)\|_{H^{-t}(\Omega)} \leq C_G(\max\{\|v\|_{H^t(\Omega)}, \|w\|_{H^t(\Omega)}\})\|v - w\|_{H^t(\Omega)} \tag{1.15}$$

holds for all $v, w \in H_0^t(\Omega)$. In particular, for each linear $\mathcal{L} : H_0^t(\Omega) \times H^{-t}(\Omega)$ corresponding to a bounded and elliptic bilinear form and $f \in H^{-t}(\Omega)$, the problem

$$\mathcal{F}(u) := \mathcal{L}u + \mathcal{G}(u) = f \tag{1.16}$$

has a unique solution $u \in H_0^t(\Omega)$. Moreover, $\mathcal{F}$ has a continuous Fréchet derivative $D\mathcal{F}$ and there exists a neighbourhood U of u such that for $v \in U$ the Fréchet derivative $D\mathcal{F}(v)$ is an isomorphism from $H_0^t(\Omega)$ to $H^{-t}(\Omega)$, which means that

$$c\|w\|_{H^t(\Omega)} \leq \|D\mathcal{F}(v)(w)\|_{H^{-t}(\Omega)} \leq C_G(\|v\|_{H^t(\Omega)})\|w\|_{H^t(\Omega)} \qquad (1.17)$$

holds uniformly in $w \in H_0^t(\Omega)$.

Proof. For the proof that $\mathcal{G}$ maps $H_0^t(\Omega)$ into $H^{-t}(\Omega)$ and the stability property (1.15), we refer to [28, Prop. 4.1]. To show existence and uniqueness, we proceed as in [27, Remark 3.1] and [76, Lemma 1.8]. We assume that $\mathcal{G}(0) = 0$. Otherwise, the problem (1.16) can simply be modified to $\mathcal{L}u + (\mathcal{G}(u) - \mathcal{G}(0)) = f - \mathcal{G}(0)$. Because G is monotone, even $\mathcal{G}$ is monotone. Thus, for all $v \in H_0^t(\Omega)$, it holds that

$$\langle \mathcal{G}(v), v \rangle = \langle \mathcal{G}(v) - \mathcal{G}(0), v - 0 \rangle \geq 0.$$

Hence, with $\mathcal{L}$ being elliptic, we conclude that

$$\langle \mathcal{F}(v), v \rangle = \langle \mathcal{L}v, v \rangle + \langle \mathcal{G}(v), v \rangle \geq \langle \mathcal{L}v, v \rangle \gtrsim \|v\|_{H^t(\Omega)}^2. \qquad (1.18)$$

Moreover, due to the ellipticity of $\mathcal{L}$, the monotonicity carries over to $\mathcal{F}$,

$$\langle \mathcal{F}(v) - \mathcal{F}(v), v - w \rangle = \langle \mathcal{L}(v - w), v - w \rangle + \langle \mathcal{G}(v) - \mathcal{G}(v), v - w \rangle \geq 0 \qquad (1.19)$$

In addition, from (1.15) and the boundedness of $\mathcal{L}$, we have

$$\|\mathcal{F}(v) - \mathcal{F}(w)\|_{H^{-t}(\Omega)} \leq (\|\mathcal{L}\| + C_G(\max\{\|v\|_{H^t(\Omega)}, \|w\|_{H^t(\Omega)}\}))\|v - w\|_{H^t(\Omega)} \qquad (1.20)$$

Because $\mathcal{F}$ fulfills (1.18), is continuous in the sense of (1.20) and monotone, see (1.19), the Browder-Minty-Theorem, see [98, Th. 10.49], gives us the existence of a solution $u \in H_0^t(\Omega)$ to (1.16). To see that the solution is unique, let $u, \tilde{u} \in H_0^t(\Omega)$ be two such solutions. Then, by the monotonicity of $\mathcal{G}$ and the ellipticity of $\mathcal{L}$, it holds that

$$0 = \langle \mathcal{F}(u) - \mathcal{F}(\tilde{u}), u - \tilde{u} \rangle \geq \langle \mathcal{L}(u - \tilde{u}), u - \tilde{u} \rangle \gtrsim \|u - \tilde{u}\|_{H^t(\Omega)},$$

hence $u = \tilde{u}$. For the proof of (1.17), we refer again to [27, Remark 3.1]. $\qquad \square$

1.3 Examples

Let us now give two examples of partial differential equations that can be treated within our setting.

Example 1.3.1. *As a first standard problem, see also [27], we return to the introductory example*

$$-\Delta u + u^3 = f \ \text{in } \Omega,$$
$$u = 0 \ \text{on } \partial\Omega,$$

where $\Delta := \sum_{i=1}^{d} \frac{\partial^2}{\partial x_i^2}$ *is the Laplace operator and* $f \in H^{-1}(\Omega)$. *Note that the operator* $-\Delta$ *can be written in the form* (1.9) *with* $t = 1$ *and* $\alpha_{e_i,e_i} = 1$, *where* e_i *is the i-th unit vector. All other coefficients vanish in this example. Therefore, the operator* L *in* (1.9) *is uniformly elliptic and by Theorem 1.2.1, the induced bilinear form*

$$a(u,v) = \int_{\Omega} \nabla u \cdot \nabla v \, \mathrm{d}x,$$

where $\nabla u \cdot \nabla v$ *is to be read as the scalar product between the gradient vectors, is bounded and elliptic on* $H_0^1(\Omega) \times H_0^1(\Omega)$. *Moreover, it is clear that the function* $G(u) = u^3$ *fulfills Estimate* (1.14) *with* $p = 3$ *and for all* $n^* \in \mathbb{N}$. *Hence, using the condition* $p \leq \frac{d+2t}{d-2t}$ *from Theorem 1.2.2, we see that for* $d \leq 3$, *the weak form*

$$a(u,v) + \langle \mathcal{G}(u), v \rangle_{H^{-1}(\Omega) \times H_0^1(\Omega)} = \langle f, v \rangle_{H^{-1}(\Omega) \times H_0^1(\Omega)}, \ v \in H_0^1(\Omega)$$

has a unique solution $u \in H_0^1(\Omega)$. *Although not covered by Theorem 1.2.2, this result also holds in the case* $d = 4$, *see [76, Example 1.11].*

Example 1.3.2. *The general setting also allows for the treatment of higher order equations. Therefore, let us consider the* biharmonic equation *with a nonlinear term*

$$\Delta^2 u + \varepsilon \sin(u) = f \ in \ \Omega,$$

$$u = \frac{\partial u}{\partial n} = 0 \ on \ \partial\Omega.$$

This is an equation of order $2t = 4$. *Because of* $\Delta^2 = \sum_{i=1}^{d} \sum_{j=1}^{d} \frac{\partial^2}{\partial x_i^2} \frac{\partial^2}{\partial x_j^2}$, *it is* $a_{2e_i,2e_j} = 1$ *for all* $1 \leq i,j \leq d$. *Therefore, we have*

$$\sum_{|\alpha|=|\beta|=2} a_{\alpha,\beta} \xi^{\alpha+\beta} = (\xi_1^2 + \ldots + \xi_d^2)^2 = |\xi|^4,$$

hence $L = \Delta^2$ *is uniformly elliptic, and by Theorem 1.2.1, the corresponding bilinear form*

$$a(u,v) = \int_{\Omega} \Delta u \cdot \Delta v \, \mathrm{d}x$$

is bounded and elliptic on $H_0^2(\Omega) \times H_0^2(\Omega)$. *Moreover, it holds that*

$$\|\sin(u)\|_{H^{-2}(\Omega)} = \sup_{\substack{v \in H_0^2(\Omega) \\ \|v\|_{H^2(\Omega)}=1}} \int_{\Omega} \sin(u(x))v(x) \, \mathrm{d}x$$

$$\leq \sup_{\substack{v \in H_0^2(\Omega) \\ \|v\|_{H^2(\Omega)}=1}} \|\sin(u)\|_{L_2(\Omega)} \|v\|_{L_2(\Omega)}$$

$$\lesssim \|u\|_{L_2(\Omega)} \lesssim \|u\|_{H^2(\Omega)}.$$

Because the energy norm induced by $a(\cdot, \cdot)$ is equivalent to the Sobolev norm, the nonlinearity $\mathcal{G}(u) := \varepsilon \sin(u)$ is indeed a contraction for sufficiently small values of ε. Therefore, by Lemma 1.0.1, there exists a unique solution $u \in H_0^2(\Omega)$ for the weak formulation of the above equation.

1.4 The stationary Navier-Stokes equation

In this section, we present a prominent example from fluid dynamics, which, at least at first glance, does not immediately fit into the setting described so far. However, we will see that, with a few modifications, it can be treated with similar techniques as the examples described above. The homogeneous, stationary, incompressible *Navier-Stokes equation* is given by

$$-\frac{1}{\text{Re}}\Delta u + (u \cdot \nabla)u + \nabla p = f \text{ in } \Omega, \tag{1.21}$$

$$\text{div } u = 0 \text{ in } \Omega, \tag{1.22}$$

$$u = 0 \text{ on } \partial\Omega. \tag{1.23}$$

Here, $u = (u_1, \dots, u_d) : \Omega \to \mathbb{R}^d$, $d \geq 2$, describes the unknown velocity of a fluid in a bounded Lipschitz domain Ω under the given external force f, and p is an unknown pressure term. Since the pressure term only appears in terms of its gradient, it is typically assumed to be normalized so that $\int_\Omega p(x) \, dx = 0$. The fixed number $\text{Re} > 0$ is the *Reynolds number* that describes the viscosity of the fluid. The divergence operator is defined by $\text{div } u = \sum_{i=1}^d \frac{\partial u_i}{\partial x_i}$ and the nonlinear expression $(u \cdot \nabla)u$ reads as $(u \cdot \nabla)u = \sum_{i=1}^d u_i \frac{\partial u}{\partial x_i}$. The linearized version of this equation, where (1.21) is replaced by

$$-\Delta u + \nabla p = f \text{ in } \Omega$$

and (1.22) and (1.23) are kept, is known as the *Stokes equation* with Dirichlet boundary conditions. Although it is not immediately clear how these equations can be treated within the general setting described above, the *LeRay* weak formulation we will explain now leads to a more familiar form. Details on the weak formulation can, for instance, be found in the standard reference [114].

Let $v = (v_1, \dots, v_d) : \Omega \to \mathbb{R}^d$ be a smooth, divergence-free function, $\text{div } v = 0$, with $v = 0$ on $\partial\Omega$. If this function is integrated against Equation (1.21), and u and p are smooth, by partial integration we have

$$\frac{1}{\text{Re}}\sum_{i=1}^d \int_\Omega \nabla u_i \nabla v_i \, dx + \int_\Omega v \cdot (u \cdot \nabla)u \, dx + \int_\Omega \nabla p \cdot v \, dx = \int_\Omega f \cdot v \, dx.$$

Now, because of $\int_\Omega \nabla p \cdot v \, dx = \int_\Omega p \, \text{div } v \, dx = 0$, the pressure term drops out. Hence, we define

$$a(v, w) := \sum_{i=1}^{d} \int_{\Omega} \nabla v_i \nabla w_i \, dx, \tag{1.24}$$

set $f(v) := \int_{\Omega} f \cdot v \, dx$ and test in the space

$$V := V(\Omega) := \{v \in (H_0^1(\Omega))^d, \text{ div } v = 0\}.$$

The latter space will be equipped with the norm on $(H^1(\Omega))^d$ and the dual space, denoted by $V' = (V(\Omega))'$, will be provided with the canonical operator norm. Then, a weak formulation of the Navier-Stokes equation is given by:

Find a $u \in V$ such that for all $v \in V$, we have

$$a(u, v) + \text{Re} \int_{\Omega} v \cdot (u \cdot \nabla) u \, dx = \text{Re} \cdot f(v). \tag{1.25}$$

Although the nonlinear term takes a more complicated, quadratic form than in the previous examples, the weak form we have arrived at is similar to those treated before. To understand the properties of the nonlinear term and to see that it is well-defined and continuous, the following lemma is essential. In similar form, the lemma and its proof stem from [86] and [114].

Lemma 1.4.1. *For $d \leq 4$, the trilinear form*

$$b(u, v, w) := \int_{\Omega} u \cdot (v \cdot \nabla) w \, dx \tag{1.26}$$

is bounded on $H_0^1(\Omega)^d \times H_0^1(\Omega)^d \times H_0^1(\Omega)^d$, and thereby also on $V(\Omega) \times V(\Omega) \times V(\Omega)$. Moreover, for each $w, y \in V$, the equation

$$a(z, v) = b(v, w, y) = \int_{\Omega} v \cdot (w \cdot \nabla) y \, dx, \quad v \in V.$$

has a unique solution $z \in V$ that depends continuously on $w, y \in V$, i.e., there exists a constant C_{NS} such that

$$\|z\| \leq C_{NS} \|w\| \cdot \|y\|$$

in the energy norm $\|\cdot\| := a(\cdot, \cdot)^{1/2}$.

Proof. It is $b(u, v, w) = \sum_{i,j=1}^{d} \int_{\Omega} u_j v_i \frac{\partial w_j}{\partial x_i} \, dx$. Therefore, it is sufficient to estimate the terms $\int_{\Omega} |u_j v_i \frac{\partial w_j}{\partial x_i}|$. By the generalized Hölder inequality (1.5), it is

$$\int_{\Omega} |u_j v_i \frac{\partial w_j}{\partial x_i}| \leq \|u_j\|_{L_4(\Omega)} \|v_i\|_{L_4(\Omega)} \|\frac{\partial w_j}{\partial x_i}\|_{L_2(\Omega)} \leq \|u_j\|_{L_4(\Omega)} \|v_i\|_{L_4(\Omega)} \|w_j\|_{H^1(\Omega)}.$$

By the Sobolev inequality (1.7) and nestedness of the L_p-spaces, it is $H_0^1(\Omega) \subset H^1(\Omega) \hookrightarrow L_4(\Omega)$ for $d \in \{2, 3, 4\}$. This shows the boundedness of $b(\cdot, \cdot, \cdot)$. Moreover, the second part follows from $a(\cdot, \cdot)$ being elliptic, which means that the operator $\mathcal{L} : V(\Omega) \to V'(\Omega)$, $\mathcal{L}(v)(w) = a(v, w)$ is boundedly invertible. Combined with the boundedness of $b(\cdot, \cdot, \cdot)$, this completes the proof. $\qquad\square$

So far, we have not yet discussed whether or not the LeRay weak form (1.25) has a unique solution. Unfortunately, without further assumptions on f or Re, we cannot expect this to be true, see [114, Ch. II]. In our methods, however, we will be concerned with the easier case of small Reynolds numbers. Under these assumptions, we have the following existence and uniqueness result from [114, Ch. II, Th. 1.3]. To see that the definition of the spaces $V(\Omega)$ from there coincides with our definition, see [114, Ch. I, Th. 1.6].

Theorem 1.4.2. *Let $d \leq 4$. There exists a constant c_{NS} only dependent on d and Ω such that if*

$$\frac{1}{\text{Re}^2} > c_{NS}\|f\|_{V'},$$

then the LeRay weak form (1.25) of the stationary, incompressible Navier-Stokes equation has a unique solution $u \in V$.

To sum up, we have seen that an adequate weak formulation leads to quite a similar problem as for standard semilinear elliptic equations. Therefore, there is hope that it can be tackled with similar techniques. However, it is already obvious now that in order to work with this particular weak form, a generating system for V is essential. This will be discussed in Section 2.2.6.

Chapter 2

Wavelet Bases and Frames

In this chapter, we outline how the equations presented in the previous chapter can be discretized, by which we understand that they are being transformed into equivalent equations in a sequence space. To do so, we make use of wavelet bases and frames. The basic idea of such wavelets is to construct bases for function spaces by translation and dilation of a single or very few functions. In particular, we work with compactly supported spline wavelets. Originally constructed in $L_2(\Omega)$, these wavelets characterize a whole range of spaces including the Sobolev spaces arising in the weak formulation of elliptic equations. Moreover, when used for the discretization of these equations, they lead to near-sparse systems that allow for an efficient numerical treatment. In addition, due to the spline structure, the computation of integrals or the pointwise evaluation of such wavelets is fast and reliable.

In contrast to bases, frames are possibly overcomplete generating systems for a Hilbert space. In the case of Sobolev spaces on bounded domains, they can be constructed via an overlapping domain decomposition approach. This approach retains the desirable properties of wavelet bases while granting more flexibility in the construction. This will help us to circumvent difficulties that may arise in the construction of wavelet basis on non-convex domains. Moreover, we sketch how divergence-free wavelets can be designed that allow for the efficient numerical treatment of problems in fluid dynamics.

2.1 Riesz bases and frames in Hilbert spaces

We start by introducing the concepts of *Riesz bases* and *frames*, respectively, first for some seperable Hilbert space $\mathcal{H}$. The concrete constructions for the Lebesgue and Sobolev spaces appearing in the weak formulation of an elliptic partial differential equation will be described in the subsequent sections. In this section, we roughly follow the lines from [18] and [104], where proofs to the statements given here can be found.

2.1.1 Riesz bases

To define a Riesz basis, let us first recall that for a countable index set $\mathcal{N}$, the sequence spaces $\ell_p(\mathcal{N})$, $0 \leq p \leq \infty$ are defined as the spaces of all sequences $\mathbf{c} = (c_n)_{n \in \mathcal{N}}$ such that

$$\|\mathbf{c}\|_{\ell_p(\mathcal{N})} := \begin{cases} \left(\sum_{n \in \mathcal{N}} |c_n|^p \right)^{1/p} & , \quad p < \infty, \\ \sup_{n \in \mathcal{N}} |c_n| & , \quad p = \infty, \end{cases}$$

is finite. The spaces are Banach spaces for $p \geq 1$ and quasi-Banach-spaces for $0 < p < 1$. The particular space $\ell_2(\mathcal{N})$ is even a Hilbert space with the inner product $\langle \mathbf{c}, \mathbf{d} \rangle_{\ell_2(\mathcal{N})} := \sum_{n \in \mathcal{N}} c_n \overline{d_n}$. In this thesis, we work with real-valued sequences only. A Riesz basis for a Hilbert space $\mathcal{H}$ can now be defined as a generating system such that the norm of an element $f \in \mathcal{H}$ is equivalent to the ℓ_2-norm of its expansion coefficients.

Definition 2.1.1. *A sequence $\{f_n\}_{n \in \mathcal{N}} \subset \mathcal{H}$ is called a* Riesz basis *for $\mathcal{H}$ if*

$$\mathrm{clos}_{\mathcal{H}} \mathrm{span}\{f_n, \, n \in \mathcal{N}\} = \mathcal{H}$$

and if there exist constants $A, B > 0$ such that

$$A\|\mathbf{c}\|_{\ell_2(\mathcal{N})}^2 \leq \| \sum_{n \in \mathcal{N}} c_n f_n \|_{\mathcal{H}}^2 \leq B\|\mathbf{c}\|_{\ell_2(\mathcal{N})}^2 \tag{2.1}$$

holds uniformly in $\mathbf{c} \in \ell_2(\mathcal{N})$. The ambiguous numbers A, B are then called the Riesz constants *of the basis $\{f_n\}_{n \in \mathcal{N}}$.*

If $\mathcal{R} = \{f_n\}_{n \in \mathcal{N}}$ is a Riesz basis for $\mathcal{H}$, then every element $f \in \mathcal{H}$ has a unique expansion $f = \mathbf{c}^\top \mathcal{R}$ with $\mathbf{c} \in \ell_2(\mathcal{N})$, where $\mathbf{c}^\top \mathcal{R}$ is the shorthand notation for $\sum_{n \in \mathcal{N}} c_n f_n$. However, it is not directly clear how to calculate the expansion coefficients c_n. At least in theory, one way to cope with this problem is to make use of a *dual Riesz basis*, the existence of which is guaranteed by the following theorem.

Theorem 2.1.2 ([18, Th. 3.6.3]). *Let $\{f_n\}_{n \in \mathcal{N}}$ be a Riesz basis for $\mathcal{H}$. Then, there exists a unique sequence $\{g_n\}_{n \in \mathcal{N}}$ in $\mathcal{H}'$ such that*

$$f = \sum_{n \in \mathcal{N}} \langle f, g_n \rangle f_n$$

holds for all $f \in \mathcal{H}$. This sequence $\{g_n\}_{n \in \mathcal{N}}$ is a Riesz basis for $\mathcal{H}'$, called dual Riesz basis. *The bases $\{f_n\}_{n \in \mathcal{N}}$ and $\{g_n\}_{n \in \mathcal{N}}$ are* biorthogonal,

$$\langle f_k, g_n \rangle = \delta_{kn}.$$

Let us remark that the dual Riesz basis can also be considered as a basis for $\mathcal{H}$ instead of $\mathcal{H}'$. Via the Riesz isomorphism, see, e.g., [79, Section 3.8], this is an equivalent description that is often found in the literature. In our applications, however, it will be more natural to explicitly work with the dual space.

A Riesz basis $\mathcal{F} = \{f_n\}_{n \in \mathcal{N}}$ allows every element in the Hilbert space to be represented in the basis with *unique* coefficients in $\ell_2(\mathcal{N})$. More precisely, the operator

$$F^* : \ell_2(\mathcal{N}) \to \mathcal{H}, \quad \mathbf{c} \mapsto \mathbf{c}^\top \mathcal{F} = \sum_{n \in \mathcal{N}} c_n f_n. \tag{2.2}$$

is well-defined and boundedly invertible.

2.1.2 Hilbert frames

A different, possibly more flexible, approach is to give up the uniqueness of the representation and instead to allow the generating system to be *overcomplete*. This will be covered by the concept of a *frame*.

Definition 2.1.3. *A sequence $\{f_n\}_{n \in \mathcal{N}} \subset \mathcal{H}$ is called a* (Hilbert) frame *for $\mathcal{H}$, if there exist constants $A, B > 0$ such that*

$$A\|f\|_{\mathcal{H}'}^2 \leq \|(\langle f, f_n \rangle_{\mathcal{H}' \times \mathcal{H}})_{n \in \mathcal{N}}\|_{\ell_2(\mathcal{N})}^2 \leq B\|f\|_{\mathcal{H}'}^2$$

holds for all $f \in \mathcal{H}'$. The constants A, B are called frame constants.

Via the mapping F^*, an equivalent, maybe more intuitive, characterization can be shown.

Theorem 2.1.4 ([18, Th. 5.5.1]). *A sequence $\{f_n\}_{n \in \mathcal{N}} \subset \mathcal{H}$ is a frame for $\mathcal{H}$ if and only if the operator F^* from (2.2) is well-defined and surjective.*

It is important to note that the operator F^* is not necessarily injective, hence the representation of an element in $\mathcal{H}$ as a linear combination of the frame elements is not unique in general. The operator F^* corresponding to a frame is called *synthesis operator*. Its dual operator, the *analysis operator*, is given by

$$F : \mathcal{H}' \to \ell_2(\mathcal{N}), \, f \mapsto (\langle f, f_n \rangle_{\mathcal{H}' \times \mathcal{H}})_{n \in \mathcal{N}}. \tag{2.3}$$

Both F and F^* are bounded, F is injective and we have the decomposition

$$\ell_2(\mathcal{N}) = \ker F^* \oplus \operatorname{ran} F.$$

The composition $S := F^* F$ is called *frame operator*. This operator is boundedly invertible and the sequence $S^{-1} \mathcal{F} := \{S^{-1} f_n\}_{n \in \mathcal{N}}$ is a frame for $\mathcal{H}'$, called the *canonical dual frame*. Moreover, with the help of this canonical dual, we obtain the *reconstruction formula*

$$g = \sum_{n \in \mathcal{N}} \langle g, S^{-1} f_n \rangle f_n, \quad g \in \mathcal{H},$$

see [18, Section 5.1]. The reconstruction coefficients $(\langle g, S^{-1} f_n \rangle)_{n \in \mathcal{N}}$ are those with minimal ℓ_2-norm among all sequences $\mathbf{c} \in \ell_2(\mathcal{N})$ with $g = \mathbf{c}^\top \mathcal{F}$. In other words, the orthogonal projector onto $\operatorname{ran} F$ is given by $\mathbf{Q} := FS^{-1}F^*$. Unfortunately, in practical applications, the computation of the canonical dual frame or the inversion of the frame operator is often difficult and additional desirable properties of the original frame elements f_n might get lost upon applying the inverse frame operator. For instance, there are frames for $L_2(\mathbb{R})$ consisting of functions in C^k for arbitrary $k \in \mathbb{N}$, where the elements of the dual frame are not even continuous, compare [48, Section 3.3]. Therefore, besides from the canonical dual frame, it can make sense to consider other *dual frames*. These are frames $\{g_n\}_{n \in \mathcal{N}}$ for $\mathcal{H}'$ such that an analogous reconstruction formula, namely

$$g = \sum_{n \in \mathcal{N}} \langle g, g_n \rangle f_n, \quad g \in \mathcal{H}, \tag{2.4}$$

holds. If the frame is overcomplete, i.e., if it is not a Riesz basis, there are other dual frames apart from the canonical dual frame, so-called *non-canonical dual frames*. For further reading on dual frames, see [18, Section 6.1].

Let us shortly comment on the relation between frames and Riesz bases. From Theorem 2.1.4, we see immediately that every Riesz basis is a frame. The converse, however, is not true in general. For instance, the union of two orthonormal bases clearly is a frame, but not a Riesz basis. Hence, to ensure that a frame is a Riesz basis, additional criteria have to be fulfilled. One possible characterization is given by the following lemma, see [18, Th. 6.11] for the proof and further equivalent criteria.

Lemma 2.1.5. *Let $\{f_n\}_{n \in \mathcal{N}}$ be a frame for $\mathcal{H}$. Then, $\{f_n\}_{n \in \mathcal{N}}$ is a Riesz basis for $\mathcal{H}$ if and only if it has a* biorthogonal *sequence, i.e., a sequence $\{g_n\}_{n \in \mathcal{N}} \subset \mathcal{H}'$ with $\langle g_n, f_m \rangle_{\mathcal{H}' \times \mathcal{H}} = \delta_{n,m}, n, m \in \mathcal{H}.$*

Finally, we are going to give another equivalent characterization of a frame that is more similar to the definition of a Riesz bases. Recall that a Riesz bases was defined as a generating system where the ℓ_2-norm of the expansion coefficients is equivalent to the norm in $\mathcal{H}$. A similar characterization holds for frames as well, when among the possibly infinitely many representations of $f \in \mathcal{H}$ one with *smallest possible* ℓ_2-norm is considered. Remind that such a representation can be obtained via the canonical dual frame.

Lemma 2.1.6 ([123, Prop. 2.2]). *The sequence $\mathcal{F} = \{f_n\}_{n \in \mathcal{N}}$ is a frame for $\mathcal{H}$ if and only if $\operatorname{clos}_{\mathcal{H}} \operatorname{span} \mathcal{F} = \mathcal{H}$ and*

$$\|f\|_{\mathcal{H}} \approx \inf_{\mathbf{c} \in \ell_2(\mathcal{N}), \, \mathbf{c}^\top \mathcal{F} = f} \|\mathbf{c}\|_{\ell_2(\mathcal{N})}. \tag{2.5}$$

2.1.3 Gelfand frames

In our context, it will be useful to consider frames that are tailored to the Gelfand triple $H_0^s(\Omega) \subset L_2(\Omega) \subset H^{-s}(\Omega)$. These frames will first be constructed in $L_2(\Omega)$, and a properly scaled version will constitute a frame for $H_0^s(\Omega)$ or $H^{-s}(\Omega)$, respectively. To describe this mechanism in an abstract fashion, it has turned out that the concept of *Gelfand frames* as presented in [34] is the right choice. Such Gelfand frames are frames tailored to the setting of a Gelfand triple $(\mathcal{B}, \mathcal{H}, \mathcal{B}')$. Since, in our case, both $\mathcal{B} = H_0^s(\Omega)$ and $\mathcal{H} = L_2(\Omega)$ are Hilbert spaces, it will be sufficient to restrict the description to the case of Gelfand triples of Hilbert spaces.

Definition 2.1.7. *Let $(\mathcal{B}, \mathcal{H}, \mathcal{B}')$ be a Gelfand triple, where $\mathcal{B}$ and $\mathcal{H}$ are Hilbert spaces. A frame $\mathcal{F} = \{f_n\}_{n \in \mathcal{N}}$ for $\mathcal{H}$ together with a dual frame $\tilde{\mathcal{F}} = \{\tilde{f}_n\}_{n \in \mathcal{N}}$ is called a* Gelfand frame *for the above Gelfand triple, if $\mathcal{F} \subset \mathcal{B}$, $\tilde{\mathcal{F}} \subset \mathcal{B}'$ and if there exists another Gelfand triple $(\mathcal{B}_d, \ell_2(\mathcal{N}), \mathcal{B}'_d)$ of sequence spaces over $\mathcal{N}$ such that*

$$F^* : \mathcal{B}_d \to \mathcal{B}, \, \mathbf{c} \mapsto \mathbf{c}^\top \mathcal{F}, \quad \tilde{F} : \mathcal{B} \to \mathcal{B}_d, \, f \mapsto (\langle f, \tilde{f}_n \rangle_{\mathcal{B} \times \mathcal{B}'})_{n \in \mathcal{N}}$$

are bounded operators.

In the case of Gelfand frames, we have a similar reconstruction formula

$$f = \sum_{n \in \mathcal{N}} \langle f, \tilde{f}_n \rangle_{\mathcal{B} \times \mathcal{B}'} f_n, \quad f \in \mathcal{B}.$$

Typically, the sequence spaces $\mathcal{B}_d$ are *weighted ℓ_p-spaces*. For a given weight factor $\mathbf{W} := \mathrm{diag}(w_n)_{n \in \mathcal{N}}$ with $w_n > 0$, these spaces $\ell_{p,\mathbf{W}}(\mathcal{N})$, $1 \leq p \leq \infty$, are defined as the sets of all sequences $\mathbf{c} = (c_n)_{n \in \mathcal{N}}$ such that the norm

$$\|\mathbf{c}\|_{\ell_{p,\mathbf{W}}(\mathcal{N})} := \|\mathbf{W}\mathbf{c}\|_{\ell_p(\mathcal{N})}$$

is finite. In this case, an isomorphism between the sequence spaces $\ell_{p,\mathbf{W}}(\mathcal{N})$ and $\ell_p(\mathcal{N})$ or between $\ell_p(\mathcal{N})$ and $\ell_{p,\mathbf{W}^{-1}}(\mathcal{N})$ is simply given by multiplication with $\mathbf{W}$, see [97]. The relation between Gelfand frames and classical Hilbert frames is clarified by the following result.

Proposition 2.1.8 ([97, Prop. 2.3]). *Let $\mathcal{F} = \{f_n\}_{n \in \mathcal{N}}$ be a Gelfand frame for $(\mathcal{B}, \mathcal{H}, \mathcal{B}')$ with dual $\tilde{\mathcal{F}} = \{\tilde{f}_n\}_{n \in \mathcal{N}}$. Assume that the corresponding sequence spaces are given by $(\ell_{2,\mathbf{W}}(\mathcal{N}), \ell_2(\mathcal{N}), \ell_{2,\mathbf{W}^{-1}}(\mathcal{N}))$ with positive weight $\mathbf{W} > 0$. Then, $\mathbf{W}\mathcal{F} = \{w_n f_n\}_{n \in \mathcal{N}}$ and $\mathbf{W}^{-1}\tilde{\mathcal{F}} = \{w_n^{-1} \tilde{f}_n\}_{n \in \mathcal{N}}$ are Hilbert frames for $\mathcal{B}'$ and $\mathcal{B}$, respectively.*

2.2 Wavelet Riesz bases and wavelet frames

We now outline the construction principles for wavelet bases and frames. Although in the course of the thesis, we will work with frames, wavelet bases on subdomains will occur as building blocks in their construction. Hence, we start by explaining the design of wavelet bases.

Even though the construction of such bases is not the focus of this work, their properties will play a vital role in the convergence and optimality analysis. In particular, their compression properties when applied for the discretization of differential operators will be essential. Therefore, we will spend some time describing their construction. For a general introduction to wavelets, we also refer to [23, 48, 90, 125].

2.2.1 Construction principles on the real line

The most common approach to the construction of wavelet bases is via *multiresolution analysis*, first introduced for wavelet bases for $L_2(\mathbb{R})$ in [89]. The situation on the real line will help us to explain most clearly the basic ideas. With a modified approach, it is possible to adapt the construction to wavelet bases on suitable bounded domains that fulfill prescribed boundary conditions. In addition, the wavelets arising from these constructions characterize a range of classical smoothness spaces including Sobolev spaces. This means that we have at hand norm equivalences of Sobolev norms with weighted sequence norms of the expansion coefficients. This property is crucial for the application to partial differential equations. For practical constructions involving spline wavelets, we refer to the overview in Subsection 2.2.3, or to the original works [45] and [95]. The latter is the construction that will also be applied in our numerical experiments. Therefore, the following outline of the design principles will also be oriented on [95].

Let us start with the outline of the construction for $L_2(\mathbb{R})$. In this situation, a function $\psi \in L_2(\mathbb{R})$ is called an *orthogonal wavelet*, if the set

$$\Psi^{L_2(\mathbb{R})} := \{\psi_{j,k}(x) := 2^{j/2}\psi(2^j x - k),\ j,k \in \mathbb{Z}\} \tag{2.6}$$

is an orthonormal basis for $L_2(\mathbb{R})$. In this thesis, wavelet bases will be denoted by Ψ. Since we will make use of wavelet bases for other spaces as well, the corresponding space will often be added as a superscript, which is omitted if it is clear from the context. The simplest example for an orthogonal wavelet in $L_2(\mathbb{R})$ is the *Haar wavelet*

$$\psi(x) := \begin{cases} 1 & x \in [0, \frac{1}{2}) \\ -1 & x \in [\frac{1}{2}, 1) \\ 0 & \text{else} \end{cases} . \tag{2.7}$$

In numerical applications, however, additional symmetry or smoothness properties are often of vital importance. Unfortunately, it turns out that the condition of orthogonality is too restrictive if such requirements are to be fulfilled, see, e.g., [48]. A less restrictive definition is to call a function ψ a *biorthogonal wavelet*, if $\Psi^{L_2(\mathbb{R})}$ as defined in (2.6) is a Riesz basis for $L_2(\mathbb{R})$ and if the dual Riesz basis $\tilde{\Psi}^{L_2(\mathbb{R})}$ is also generated by a single function $\tilde{\psi} \in L_2(\mathbb{R})$, i.e., $\tilde{\Psi}^{L_2(\mathbb{R})} = \{\tilde{\psi}_{j,k},\ j,k \in \mathbb{Z}\}$. We now introduce the most common tool for the construction of such wavelets.

Definition 2.2.1. *Let* $\phi \in L_2(\mathbb{R})$, $\phi_{j,k}(x) = 2^{j/2}\phi(2^j x - k)$ *and assume that the spaces*

$$V_j := \text{clos}_{L_2(\mathbb{R})} \text{span}\{\phi_{j,k},\ k \in \mathbb{Z}\}$$

(i) are nested, $V_{j-1} \subset V_j,\ j \in \mathbb{Z}$,

(ii) $\bigcup_{j\in\mathbb{Z}} V_j$ *is dense in* $L_2(\mathbb{R})$,

(iii) $\bigcap_{j\in\mathbb{Z}} V_j = \{0\}$,

(iv) and $\{\phi_{0,k}, k \in \mathbb{Z}\}$ is a Riesz basis for V_0.

Then, the sequence $\{V_j\}_{j \in \mathbb{Z}}$ is called a multiresolution analysis *(MRA) for $L_2(\mathbb{R})$ and ϕ is called a* scaling function *or* generator. *Two MRAs $\{V_j\}_{j \in \mathbb{Z}}$ and $\{\tilde{V}_j\}_{j \in \mathbb{Z}}$ with*

$$\tilde{V}_j := \operatorname{clos}_{L_2(\mathbb{R})} \operatorname{span}\{\tilde{\phi}_{j,k}, \, k \in \mathbb{Z}\}$$

are called biorthogonal, *if $\langle \phi, \tilde{\phi}_{0,k} \rangle_{L_2(\mathbb{R})} = \delta_{0,k}, \, j \in \mathbb{Z}$. In this case, $\{V_j\}_{j \in \mathbb{Z}}$ and $\{\tilde{V}_j\}_{j \in \mathbb{Z}}$ are called* primal *and* dual *MRA, respectively.*

In particular, from the definition of an MRA it follows that $V_0 \subset V_1$ and that $\{\phi_{1,k}, k \in \mathbb{Z}\}$ is a Riesz basis for V_1. Thus, for a scaling function ϕ there are *mask coefficients* $a_k, \, k \in \mathbb{Z}$ such that

$$\phi(x) = \sum_{k \in \mathbb{Z}} a_k \phi(2x - k).$$

A function fulfilling such a relation is also called *refinable function*. Let ϕ and $\tilde{\phi}$ be biorthogonal scaling functions with mask coefficients a_k and $\tilde{a}_k$, respectively. Moreover, assume they are normalized so that $\int_{\mathbb{R}} \phi(x) \, \mathrm{d}x = \int_{\mathbb{R}} \tilde{\phi}(x) \, \mathrm{d}x = 1$. Now define

$$\psi(x) := \sum_{k \in \mathbb{Z}} (-1)^k \tilde{a}_{1-k} \phi(2x - k), \quad \tilde{\psi}(x) := \sum_{k \in \mathbb{Z}} (-1)^k a_{1-k} \tilde{\phi}(2x - k).$$

If ϕ and $\tilde{\phi}$ are compactly supported and the masks a_k and $\tilde{a}_k$ are finite as in many examples, the functions ψ and $\tilde{\psi}$ are compactly supported as well. The construction of primal and dual scaling functions with such properties is far from trivial and typically involves Fourier techniques. Since we will later mostly be interested in bounded domains, where these techniques are not available, we do not present the details here. Instead, we refer to the literature, where a detailled discussion can for instance be found in [29]. For such constructions, it can be shown that the spaces defined by $W_j = \operatorname{span}_{L_2(\mathbb{R})}\{\psi_{j,k}, \, j, k \in \mathbb{Z}\}$ and $\tilde{W}_j = \operatorname{span}_{L_2(\mathbb{R})}\{\tilde{\psi}_{j,k}, \, j, k \in \mathbb{Z}\}$ are complements of the V_j and $\tilde{V}_j$, i.e.,

$$V_{j+1} = V_j \oplus W_j, \quad \tilde{V}_{j+1} = \tilde{V}_j \oplus \tilde{W}_j, \tag{2.8}$$

and the biorthogonality conditions

$$V_j \perp \tilde{W}_j, \quad \tilde{V}_j \perp W_j \tag{2.9}$$

hold. The spaces W_j can be considered the refinements from the coarser level V_{j-1} to the finer level V_j. In particular, it holds that

$$L_2(\mathbb{R}) = \bigoplus_{j \in \mathbb{Z}} W_j = V_{j_0} \oplus \bigoplus_{j \geq j_0} W_j$$

for any $j_0 \in \mathbb{Z}$. In the orthogonal case, $\psi = \tilde{\psi}$, these are indeed wavelets for $L_2(\mathbb{R})$. In the more general, biorthogonal case, additional conditions are necessary, see again [29]. These conditions again involve Fourier techniques that are not applicable on the interval or even more general bounded domains. Hence, the issue of stability over all refinement levels will be discussed later in the interval case that is relevant to our applications.

2.2.2 Construction principles on bounded domains

To construct wavelets that are suitable for the discretization of PDEs on bounded domains, it is necessary to adapt the concepts described above for $L_2(\mathbb{R})$ to the situation on bounded domains. Although most practical constructions are fitted to the unit interval, the theory can be described for general bounded domains. The situation on bounded domains is somewhat more complicated than on the real line, in particular if moment and boundary conditions have to be incorporated into the construction. Roughly following [44] and [94, 95], we now sketch the basic ideas for the construction. In this section, we mostly outline the general construction principles. A practical construction on the unit interval will be described afterwards. For further wavelet constructions, we also refer to [20, 45]. The first step is to adapt the concept of multiresolution analysis.

Definition 2.2.2. *Let Ω be a bounded Lipschitz domain, $j_0 \in \mathbb{Z}$, Δ_j be finite index sets for all $j \geq j_0$, and $\Phi_j = \{\phi_{j,k}, \ k \in \Delta_j\} \subset L_2(\Omega)$. The spaces*

$$V_j := \operatorname{span} \Phi_j$$

form a multiresolution analysis *for $L_2(\Omega)$ if*

(i) *they are nested, $V_{j-1} \subset V_j$,*

(ii) *they create whole space, $\operatorname{clos}_{L_2(\Omega)} \bigcup_{j \geq j_0} V_j = L_2(\Omega)$,*

(iii) *and if Φ_j and $\tilde{\Phi}_j$ are uniformly stable, i.e., they are Riesz bases for their span with Riesz constants that can be chosen independently of $j \geq j_0$.*

Contrary to the definition on the real line, it is no longer assumed that the generators $\phi_{j,k}$ stem from dilation and translation of a single function. This is necessary to allow for modifications around the boundary. In addition, a *coarsest level j_0* is fixed. To guarantee that a pair of biorthogonal systems is uniformly stable, the following lemma will be helpful:

Lemma 2.2.3 ([44, Lemma 2.1, (i)]). *Let Φ_j and $\tilde{\Phi}_j$ be biorthogonal systems such that the functions are* uniformly bounded,

$$\|\phi_{j,k}\|, \|\tilde{\phi}_{j,k}\| \lesssim 1,$$

and locally finite, *i.e., the quantities*

$$\#\{k' \in \Delta_j, \operatorname{supp} \phi_{j,k} \cap \operatorname{supp} \phi_{j,k'} \neq \emptyset\}, \quad \#\{k' \in \Delta_j, \operatorname{supp} \tilde{\phi}_{j,k} \cap \operatorname{supp} \tilde{\phi}_{j,k'} \neq \emptyset\}$$

are uniformly bounded in $j \geq j_0$ and $k \in \Delta_j$. Then $\Phi := \{\Phi_j\}_{j \geq j_0}$ and $\tilde{\Phi} := \{\tilde{\Phi}_j\}_{j \geq j_0}$ are uniformly stable

In the following, we sketch how a wavelet Riesz basis can be constructed with the help of a biorthogonal MRA. Because the spaces V_j are nested, there exist *refinement matrices* $\mathbf{M}_{j,0} \in \mathbb{R}^{\#\Delta_{j+1} \times \#\Delta_j}$ such that

$$\Phi_j = \mathbf{M}_{j,0}^\top \Phi_{j+1}.$$

Note that, contrary to the setting on the real line, the refinement relation is explicitly allowed to depend on the level j. Since the Φ_j are Riesz bases for their span with Riesz constants uniformly bounded from below and above, these matrices are uniformly bounded as well,

$$\|\mathbf{M}_{j,0}\|_{\ell_2(\Delta_j) \to \ell_2(\Delta_{j+1})} = \mathcal{O}(1).$$

The next task is to construct bases Ψ_j and $\tilde{\Psi}_j$ for the complement spaces W_j and $\tilde{W}_j$. A possible strategy was outlined in [14, 15]. There, bases for the complement spaces are computed by solving linear systems for the coefficients of the elements W_j in the generator basis for V_{j+1} and explicitly enforcing biorthogonality. However, it has turned out that, in practice, a modified approach might be very efficient and good to handle in implementations. Based on [16], this method of *stable completion* will be outlined in the following.

The aim is to find rectangular matrices $\mathbf{M}_{j,1}, \tilde{\mathbf{M}}_{j,1} \in \mathbb{R}^{\#\Delta_{j+1} \times \#\nabla_j}$, $\nabla_j := \Delta_{j+1} \setminus \Delta_j$, such that the combined square matrices

$$\mathbf{M}_j := (\mathbf{M}_{j,0}, \mathbf{M}_{j,1}), \; \tilde{\mathbf{M}}_j := (\tilde{\mathbf{M}}_{j,0}, \tilde{\mathbf{M}}_{j,1})$$

are uniformly bounded in j with inverses that are also uniformly bounded in j. The matrices $\mathbf{M}_{j,1}$ and $\tilde{\mathbf{M}}_{j,1}$ are then called *stable completion* of $\mathbf{M}_{j,0}$ and $\tilde{\mathbf{M}}_{j,0}$. With these matrices, we can define bases for the complement spaces of the V_j and $\tilde{V}_j$, namely

$$\Psi_j = \mathbf{M}_{j,1}^\top \Phi_{j+1}, \; \tilde{\Psi}_j = \tilde{\mathbf{M}}_{j,1}^\top \tilde{\Phi}_{j+1}.$$

However, this construction does not guarantee that the Ψ_j and $\tilde{\Psi}_j$ are biorthogonal. To ensure this, it will be necessary to work with a special stable completion, which, in turn, can easily be constructed from any given stable completion. This is specified by the following lemma, which is a consequence of [16, Corollary 3.1, Th. 3.3].

Lemma 2.2.4. *Let Φ_j, $\tilde{\Phi}_j$ be the generators of a pair of biorthogonal multiresolution analyses with refinement matrices $\mathbf{M}_{j,0}$ and $\tilde{\mathbf{M}}_{j,0}$. Let $\check{\mathbf{M}}_{j,1}$ be a stable completion of $\mathbf{M}_{j,0}$. Then*

$$\mathbf{M}_{j,1} := (\mathbf{I} - \mathbf{M}_{j,0}\tilde{\mathbf{M}}_{j,0}^\top)\check{\mathbf{M}}_{j,1}$$

is another stable completion. Moreover,

$$\mathbf{M}_j^{-1} = \begin{pmatrix} \tilde{\mathbf{M}}_{j,0}^\top \\ \breve{\mathbf{G}}_{j,1} \end{pmatrix} =: \begin{pmatrix} \tilde{\mathbf{M}}_{j,0}^\top \\ \tilde{\mathbf{M}}_{j,1}^\top \end{pmatrix},$$

where $\breve{\mathbf{G}}_j = \begin{pmatrix} \breve{\mathbf{G}}_{j,0} \\ \breve{\mathbf{G}}_{j,1} \end{pmatrix} =: \breve{\mathbf{M}}_j^{-1}$, *and the systems*

$$\Psi_j := \mathbf{M}_{j,1}^\top \Phi_{j+1}, \quad \tilde{\Psi}_j := \breve{\mathbf{G}}_{j,1} \tilde{\Phi}_{j+1}$$

are biorthogonal, $\langle \psi_{j,k}, \tilde{\psi}_{j',k'} \rangle = \delta_{j,j'} \delta_{k,k'}$. *In addition, we have the two-scale relation*

$$\Phi_{j+1}^\top = \Phi_j^\top \tilde{\mathbf{M}}_{j,0} + \Psi_j^\top \tilde{\mathbf{M}}_{j,1}. \tag{2.10}$$

Setting $\Psi_{j_0-1} := \Phi_{j_0}$ and $\tilde{\Psi}_{j_0-1} := \tilde{\Phi}_{j_0}$ to include one level of generators, we can now define

$$\Psi^{L_2(\Omega)} := \Phi_{j_0} \cup \bigcup_{j \geq j_0} \Psi_j = \bigcup_{j \geq j_0-1} \Psi_j$$

and, analogously,

$$\tilde{\Psi}^{L_2(\Omega)} := \tilde{\Phi}_{j_0} \cup \bigcup_{j \geq j_0} \tilde{\Psi}_j = \bigcup_{j \geq j_0-1} \tilde{\Psi}_j.$$

By this, we have a pair of biorthogonal systems,

$$\langle \Psi^{L_2(\Omega)}, \tilde{\Psi}^{L_2(\Omega)} \rangle := \big(\langle \psi_{j,k}, \tilde{\psi}_{j',k'} \rangle \big)_{j,j' \geq j_0-1, \, k \in \nabla_j, k' \in \nabla_{j'}} = \mathbf{I},$$

where ∇_{j_0-1} is to be read as Δ_{j_0}. The corresponding index sets will be written as $\Lambda^{\mathcal{I}} := \bigcup_{j \geq j_0-1} \nabla_j$, thus we write $\Psi^{L_2(\Omega)} = \{\psi_\lambda\}_{\lambda \in \Lambda^{\mathcal{I}}}$ and $\tilde{\Psi}^{L_2(\Omega)} = \{\tilde{\psi}_\lambda\}_{\lambda \in \Lambda^{\mathcal{I}}}$. With $|\lambda| := j$, we denote the level of the wavelet ψ_λ.

The systems $\{\Phi_j \cup \Psi_j\}$ are automatically L_2-stable, but it is not immediately clear whether the whole systems $\Psi^{L_2(\Omega)}$ and $\tilde{\Psi}^{L_2(\Omega)}$ form a pair of biorthogonal Riesz bases for $L_2(\Omega)$, i.e., whether Riesz stability can be ensured over all levels. To discuss this point, we need to introduce some additional notation.

For $j \geq j_0$, by biorthogonality of the systems $\Phi_j, \tilde{\Phi}_j$, we can define the projector

$$Q_j f := \langle f, \tilde{\Phi}_j \rangle \Phi_j := \sum_{k \in \Delta_j} \langle f, \tilde{\phi}_{j,k} \rangle \phi_{j,k}$$

onto V_j and, analogously, its dual projector

$$Q_j^* f := \langle f, \Phi_j \rangle \tilde{\Phi}_j := \sum_{k \in \Delta_j} \langle f, \phi_{j,k} \rangle \tilde{\phi}_{j,k}$$

onto $\tilde{V}_j$. For the sake of completeness, let $Q_{j_0-1} := 0$ and $Q_{j_0-1}^* := 0$. With the help of these projectors, every $f \in L_2(\Omega)$ has the multiscale representation

$$f = \sum_{j \geq j_0-1} (Q_{j+1} - Q_j) f. \tag{2.11}$$

In particular, with $W_j = (Q_{j+1} - Q_j)L_2(\Omega)$ and $\tilde{W}_j = (Q^*_{j+1} - Q^*_j)L_2(\Omega)$, we have a characterization of the complements of the V_j and $\tilde{V}_j$ as in (2.8) subject to the biorthogonality relation (2.9). In order to be able to show Riesz stability of the systems we construct in $H^t(\Omega)$, we have to establish a norm equivalence of the Sobolev norm with a weighted sum of the norm of the terms in (2.11). To do so, we need to assume certain properties of the underlying biorthogonal multiresolution analysis. For $S \in \{V, \tilde{V}\}$ with $V := \{V_j\}_{j \geq j_0}$ and $\tilde{V} := \{\tilde{V}_j\}_{j \geq j_0}$, an estimate of the form

$$\inf_{v_j \in S_j} \|v - v_j\|_{L_2(\Omega)} \lesssim 2^{-sj} \|v\|_{H^s(\Omega)}, \quad 0 \leq s \leq \sigma_S \tag{2.12}$$

is called *Jackson inequality*. It describes the approximation power of the underlying approximation spaces. To ensure the validity of such an estimate, it is essential that the generators can reproduce polynomials.

Lemma 2.2.5 ([44, Lemma 2.1, (ii)]). *Assume that the conditions of Lemma 2.2.3 are fulfilled and, in addition, the space Π_l of polynomials with degree at most $l - 1$ is contained in each V_j, and the primal and dual generators are locally supported,*

$$\sup_{k \in \Delta_j} \{\operatorname{diam} \operatorname{supp} \phi_{j,k}, \operatorname{diam} \operatorname{supp} \tilde{\phi}_{j,k}\} \lesssim 2^{-j}.$$

Then, it holds that

$$\inf_{v_j \in V_j} \|v - v_j\|_{L_2(\Omega)} \lesssim 2^{-lj} \|v\|_{H^l(\Omega)}.$$

Interchanging the roles of V_j and $\tilde{V}_j$, it is clear that such an estimate can also be shown for the dual multiresolution analysis if the $\tilde{V}_j$ can reproduce polynomials up to some order $\tilde{l} - 1$. However, a Jackson inequality alone is not sufficient to guarantee Riesz stability. A second ingredient is a *Bernstein estimate*

$$\|v_j\|_{H^s(\Omega)} \lesssim 2^{sj} \|v_j\|_{L_2(\Omega)}, \quad v_j \in S_j, 0 \leq s \leq \gamma_S. \tag{2.13}$$

This estimate can be verified for most constructions that are relevant to our practical applications, see [44, 45, 95]. To show results of this kind, one typically has to make use of the smoothness of primal and dual generators in the Sobolev scale. Putting together the two estimates above, the desired norm equivalence can be derived.

Theorem 2.2.6 ([40, Corollary 5.2]). *Let V and $\tilde{V}$ be biorthogonal MRAs that fulfill a Jackson estimate (2.12) for $0 \leq s \leq \sigma_V$ and $0 \leq s \leq \sigma_{\tilde{V}}$, respectively, and a Bernstein estimate (2.13) for $0 \leq s \leq \gamma_V$ and $0 \leq s \leq \gamma_{\tilde{V}}$, respectively. Let $t := \min\{\sigma_V, \gamma_V\}$ and $\tilde{t} := \min\{\sigma_{\tilde{V}}, \gamma_{\tilde{V}}\}$. Then, for $s \in (-\tilde{t}, t)$, we have*

$$\|v\|_{H^s(\Omega)} \eqsim \left(\sum_{j \geq j_0 - 1} 2^{2sj} \|(Q_{j+1} - Q_j)v\|^2_{L_2(\Omega)} \right)^{1/2}, \tag{2.14}$$

where, exceptionally, $H^s(\Omega)$ for $s < 0$ is to be read as the dual of $H^{-s}(\Omega)$.

Using that the systems $\Psi^{L_2(\Omega)}$ and $\tilde{\Psi}^{L_2(\Omega)}$ are biorthogonal, under the same assumptions and with the same notation as in Theorem 2.2.6, we have the norm equivalence

$$\|v\|_{H^s(\Omega)} \eqsim \left(\sum_{j \geq j_0-1} \sum_{k \in \nabla_j} 2^{2sj} |\langle v, \tilde{\psi}_{j,k} \rangle|^2 \right)^{1/2}, \quad s \in (-\tilde{t}, t), \tag{2.15}$$

see [44, Corollary 2.7]. In particular, with $s = 0$, we see as in loc. cit. that $\Psi^{L_2(\Omega)}$ is a Riesz basis for $L_2(\Omega)$. That is because by (2.15), $\tilde{\Psi}^{L_2(\Omega)}$ is a frame for $L_2(\Omega)$ and by Lemma 2.1.5, it is even a Riesz basis, which by Theorem 2.1.2 also holds for $\Psi^{L_2(\Omega)}$. The norm equivalence (2.15), however, holds for a whole range of Sobolev spaces with parameters s. Hence, in a next step the Riesz basis property is extended to such spaces. To do so, let us introduce the diagonal scaling matrix

$$\mathbf{D} := \operatorname{diag}(2^{|\lambda|})_{\lambda \in \Lambda^{\mathcal{I}}}.$$

With the help of this matrix, we can define the scaled version

$$\Psi^{H^s(\Omega)} := \mathbf{D}^{-s} \Psi^{L_2(\Omega)} = \{2^{-s|\lambda|} \psi_\lambda, \lambda \in \Lambda^{\mathcal{I}}\}.$$

The following proposition states that for the range of positive parameters for which (2.2.6) holds, this is indeed a Riesz basis for $H^s(\Omega)$.

Proposition 2.2.7 ([123, Prop. 2.6]). *Under the same conditions as in Theorem 2.2.6 and for $s \in [0, t)$, $\Psi^{H^s(\Omega)}$ is a Riesz basis for $H^s(\Omega)$.*

Moreover, note that the dual Riesz basis in $H^{-s}(\Omega)$ can explicitly be written down as $\{2^{s|\lambda|} \tilde{\psi}_\lambda\}_{\lambda \in \Lambda^{\mathcal{I}}}$. This directly follows from Theorem 2.1.2 and the biorthogonality relation

$$\langle 2^{-s|\lambda|} \psi_\lambda, 2^{s|\mu|} \tilde{\psi}_\mu \rangle_{H^s(\Omega) \times H^{-s}(\Omega)} = 2^{s(|\mu|-|\lambda|)} \langle \psi_\lambda, \tilde{\psi}_\mu \rangle_{L_2(\Omega)} = \delta_{\lambda,\mu}, \quad \lambda, \mu \in \Lambda^{\mathcal{I}}.$$

In summary, we have now outlined the rather abstract construction principles for wavelet bases. In the next section, we will sketch a practical construction involving spline wavelets.

2.2.3 Spline wavelets on the interval

Let us now describe the design of a spline wavelet basis on the unit interval $\mathcal{I} := [0, 1]$, following [95]. This approach is based on the construction on the real line from [29] and leads to well-conditioned bases. The details of the construction are technically rather involved, so for a full description and proofs we refer to [95] and the references therein. Spline wavelets are not only desirable from a theoretical point of view because they are piecewise polynomial and supported only on a small interval, but they are also relatively easy to handle in numerical applications. This is mostly because the pointwise evaluation and the derivatives and integrals of piecewise polynomials are easy to compute.

The primal MRA is constructed by using *Schoenberg splines* with multiple knots at the left and right end of the interval. Hence, for a given fixed desired degree $l \geq 2$ of polynomial exactness and all $j \in \mathbb{N}$, consider the knot sequence $T_l^j := \{t_k^j\}_{k=-l+1}^{2^j+l-1}$ of equidistant knots with multiplicity $l+1$ on the boundary,

$$
t_k^j = \begin{cases} 0, & -l+1 \leq k \leq 0 \\ 2^{-j}k, & 1 \leq k \leq 2^j - 1 \\ 1, & 2^j \leq k \leq 2^j + l - 1 \end{cases},
$$

and the B-splines

$$
B_{k,l}^j(x) := (t_{k+l}^j - t_k^j)[t_k^j, \ldots, t_{k+l}^j; \max\{0, t - x\}^{d-1}], \quad -l+1 \leq k \leq 2^j - 1.
$$

Recall that for a knot sequence $\{t_k\}_{k=0}^n$ the *divided difference*

$$
[t_0, \ldots, t_n; f(t)]
$$

is defined as the uniquely determined leading coefficient of the interpolation polynomial determined by the points $(t_k, f(t_k))$, $0 \leq k \leq n$. The B-splines $B_{k,l}^j$ are compactly supported in $[t_k^j, t_{k+l}^j]$ and symmetric around one half. Moreover, the simple relation $B_{k,l}^{j+1}(x) = B_{k,l}^j(2x)$ holds between B-splines on subsequent levels. We can now define the generators on level $j \in \mathbb{N}$ as

$$
\Phi_j := \{2^{j/2} B_{k,l}^j, \ -l+1 \leq k \leq 2^j - 1\}.
$$

Then, for any fixed coarsest level $j_0 \in \mathbb{N}$, the spaces $V_j = \operatorname{span} \Phi_j$, $j \geq j_0$ indeed constitute an MRA for $L_2([0,1])$ with order l of polynomial exactness. For another parameter $\tilde{l} \in \mathbb{N}$ with $\tilde{l} \geq l$ and $l + \tilde{l}$ even, a dual biorthogonal MRA can be constructed such that it has polynomial exactness of order $\tilde{l}$. The construction is rather technical and the dual generators are no longer spline functions. Fortunately, the explicit use of the dual basis is hardly needed in the applications we have in mind. However, with this construction it can be shown that both the primal and dual MRA fulfill the desired Jackson and Bernstein inequalities. After the construction of the primal and dual MRA, the wavelets are computed using the method of stable completion as outlined in the previous subsection. With this construction, the wavelets are compactly supported and, except for the functions around the boundary, identical with the wavelets from [29].

Another advantage of using the Schoenberg splines is the fact that it is relatively easy to incorporate Dirichlet boundary conditions. Such boundary conditions only have to be built into the primal generators and wavelets, whereas on the dual side, it is desirable to retain the full degree of polynomial reproduction. Doing so, the vanishing moments of the primal basis are preserved. On the primal side, this can be done by simply omitting the outermost generator functions, which are the only generators that do not vanish at the boundary. To keep the dimensions of the primal and dual multiresolution spaces equal, it is necessary to adapt the dual MRA as well. For details, we refer to [95, Section 5.3].

2.2.4 From the interval to the unit cube

For the construction of wavelet bases in multiple dimensions, it is a well-known standard approach to start with a wavelet basis on the unit interval and extend it to the unit square $\square := (0,1)^d$ by means of tensor products. As a starting point, let

$$\Psi^{L_2(\mathcal{I})} = \Phi_{j_0} \cup \bigcup_{j \geq j_0} \Psi_j$$

be a wavelet basis for $L_2(\mathcal{I})$ as constructed, for instance, in [44, 45, 95], see also the overview in the previous subsection. Roughly following the exposition from [97], we first fix some notation and then define the wavelets on the unit cube. Recall that the index set $\Lambda^{\mathcal{I}}$ corresponding to $\Psi^{\mathcal{I}}$ is given by $\Delta_{j_0} \cup \bigcup_{j \geq j_0} \nabla_j$, where the Δ_j and ∇_j are finite index sets that typically decode the location of a generator or wavelet and j stands for the level. We will assume that $\Delta_j, \nabla_j \subset \mathbb{Z}$, which is not a restriction, because these finite sets can simply be numbered. To distinguish between generators and wavelets, we introduce another parameter $e \in \{0,1\}$ and set

$$\nabla_{j,e} = \begin{cases} \Delta_j, & e = 0 \\ \nabla_j, & e = 1 \end{cases},$$

and, for $k \in \nabla_{j,e}$, we define

$$\psi_{j,e,k} := \begin{cases} \phi_{j,k}, & e = 0 \\ \psi_{j,k}, & e = 1 \end{cases}.$$

Then, with $\mathbf{0} := (0, \ldots, 0) \in \{0,1\}^d$, we can define the set $\Lambda^{\square} \subset \mathbb{Z} \times \{0,1\}^d \times \mathbb{Z}^d$ of wavelet basis indices by

$$\begin{aligned}
\Lambda^{\square} := &\{(j_0, \mathbf{0}, \mathbf{k}), \, k_i \in \nabla_{j_0,0}, \, 1 \leq i \leq d\} \cup \\
&\{(j, \mathbf{e}, \mathbf{k}), \, j \geq j_0, \, \mathbf{e} \in \{0,1\}^d \setminus \mathbf{0}, \, k_i \in \nabla_{j,e_i}, \, 1 \leq i \leq d\}.
\end{aligned}$$

The first part of this set represents tensor products only of generators on the lowest level j_0, whereas in the second part, at least in one coordinate, a true wavelet is involved. For an index $\lambda = (j, \mathbf{e}, \mathbf{k})$, the level of a multivariate wavelet is written as $|\lambda| := j$, the parameter $\mathbf{e}$ is called *type parameter* and $\mathbf{k}$ stands for the spatial location. Having defined the index set, the wavelets on the unit cube are given by

$$\psi_{\lambda}^{L_2(\square)}(x) := \prod_{i=1}^d \psi_{j,e_i,k_i}(x_i), \quad \lambda = (j, \mathbf{e}, \mathbf{k}) \in \Lambda^{\square}, \, x \in \square.$$

The collection

$$\Psi^{L_2(\square)} := \{\psi_{\lambda}^{L_2(\square)}, \, \lambda \in \Lambda^{\square}\}$$

then forms a Riesz basis for $L_2(\square)$. It inherits the properties of the underlying interval basis such as compact support, vanishing moments and norm equivalences. In particular, a scaled version $\mathbf{D}^{-s}\Psi^{L_2(\square)}$ is a Riesz basis for $H^s(\square)$ for parameters s in a range depending on the validity of Jackson and Bernstein estimates, see Theorem 2.2.6 and Proposition 2.2.7. Since Dirichlet boundary conditions can be incorporated into the construction on the interval, as laid out in the previous subsection, this also holds for the wavelet basis on the unit cube. In this case, the scaled version is a Riesz basis for $H_0^s(\square)$. The corresponding dual wavelet basis can be lifted to the unit cube in the same fashion.

A natural further step is the attempt to generalize the construction to a broader range of domains Ω. This is relatively easy if there exists a C^k-diffeomorphism, $k \geq s$, $\kappa : \square \to \Omega$ with $|\det \kappa(x)| \approx 1$. Then, for $\lambda \in \Lambda^{\square}$, the lifted wavelet is given by

$$\psi_\lambda^{L_2(\Omega)} := \frac{\psi_\lambda^{L_2(\square)}(\kappa^{-1}(x))}{|\det D\kappa(\kappa^{-1}(x))|^{\frac{1}{2}}}.$$

In this definition, the denominator is only necessary for normalization purposes. The lifted wavelets form a Riesz basis $\Psi^{L_2(\Omega)} = \{\psi_\lambda^{L_2(\Omega)}, \lambda \in \Lambda^{\square}\}$ for $L_2(\Omega)$. If the original basis $\Psi^{L_2(\square)}$ characterizes Sobolev spaces with Dirichlet boundary conditions, this also holds for the lifted version, thus a scaled version $\mathbf{D}^{-s}\Psi^{L_2(\Omega)}$ is a Riesz basis for $H_0^s(\Omega)$. Moreover, the dual Riesz basis can be lifted in the same way, see [33]. This approach clearly applies to the special case of κ being an affine, bijective mapping. Hence, for domains of the above kind such as rectangles, it is safe to assume that we have at hand suitable wavelet bases.

However, this approach cannot be applied if the domain cannot be written in the form $\Omega = \kappa(\square)$ with a diffeomorphism κ and $|\det D\kappa(x)| \approx 1$. This is, for instance, the case for the L-shaped domain

$$\Omega = (-1,1)^2 \setminus [0,1)^2, \tag{2.16}$$

which is a prototype of a two-dimensional polygonal domain with a reentrant corner. One strategy from [46] to circumvent this problem is to decompose the domain into *nonoverlapping* subdomains that, in turn, are diffeomorphic to the unit cube, see as an example Figure 2.2.4. Then, one has to construct wavelet bases for the subdomains with suitable boundary conditions. Glueing them together along the interfaces of the subdomains leads to a Riesz basis for $L_2(\Omega)$. However, the smoothness of the basis is limited in the sense that it characterizes Sobolev spaces $H^s(\Omega)$ as in (2.15) only for $s < \frac{3}{2}$. This is due to the process of glueing together wavelets from different subdomains. As we will later see in Chapters 3 and 4, this can hamper the maximal rate with which the wavelet methods we are about to construct converge. Moreover, the construction itself is rather technical and difficult to implement in practice. Hence, we will make use of a different approach from [104] using an *overlapping* domain decomposition. This approach will be sketched in the following subsection.

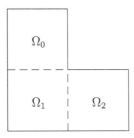

Figure 2.1: L-shaped domain with nonoverlapping domain decomposition.

2.2.5 Wavelet frames

In the previous sections, we have sketched the construction of wavelet bases on domains that are diffeomorphic images of the unit cube, but we have also pointed out the limitations of these constructions when it comes to more complicated domains such as the L-shaped domain (2.16). As an alternative, we now sketch the construction of wavelet frames with the help of an overlapping domain decomposition using the previously described wavelet bases an an ingredient in the construction. The choice of overlapping subdomains allows greater flexibility than using a nonoverlapping covering. Moreover, the problem of glueing together functions from different subdomains along the interior boundaries vanishes. The price to pay is that the resulting system is no longer a basis, so we have to deal with a redundant generating system. In the context of the numerical solution of elliptic problems, this approach was first introduced in [104]. To describe the construction in detail, a few preparations are necessary.

Definition 2.2.8. *Let Ω be a domain in $\mathbb{R}^d$. Then, $\{\Omega_i\}_{i=0}^{m-1}$ is called an* overlapping *domain decomposition of Ω, if all the Ω_i are subdomains of Ω and*

$$H_0^t(\Omega) = H_0^t(\Omega_0) + \ldots + H_0^t(\Omega_{m-1}), \quad t \in \mathbb{N}.$$

Remark 2.2.9. *In the previous definition, we have tacitly made use of the embeddings $E_i : H_0^t(\Omega_i) \to H_0^t(\Omega)$ that extend a function from $v \in H_0^t(\Omega_i)$ to a function $E_i v \in H_0^t(\Omega)$ by continuation with zero. Clearly, these extension operators are well-defined and isometric. Thereby, we often leave out the explicit use of these operators and consider a function $v \in H_0^t(\Omega_i)$ as a function on the entire domain by continuation with zero whenever necessary.*

If such an overlapping domain decomposition according to Definition 2.2.8 exists, it can moreover be shown that the decomposition is stable. This result is known as the *partition lemma*.

Lemma 2.2.10. *Let* $\mathcal{H}_0, \ldots, \mathcal{H}_{m-1}$ *be closed subspaces of a Hilbert space* $\mathcal{H}$ *such that*

$$\mathcal{H} = \mathcal{H}_0 + \ldots + \mathcal{H}_{m-1}.$$

Then, each $f \in \mathcal{H}$ *has a decomposition* $f = \sum_{i=0}^{m-1} f_i$ *with* $f_i \in \mathcal{H}_i$ *and* $\|f_i\|_{\mathcal{H}} \lesssim \|f\|_{\mathcal{H}}$.

Proof. The proof follows the lines from [88, Example 2.67]. Because the result is vital in our context, we recall the argument here. We start with the Banach space $\bar{\mathcal{H}} := \mathcal{H}_0 \times \ldots \times \mathcal{H}_{m-1}$ equipped with the norm $\|\bar{f}\|_{\bar{\mathcal{H}}} := \max_{i=0}^{m-1} \|f_i\|_{\mathcal{H}}$, where $\bar{f} = (f_0, \ldots, f_{m-1})$. Now, define the mapping

$$\eta : \bar{\mathcal{H}} \to \mathcal{H}, \quad \eta(\bar{f}) := \sum_{i=0}^{m-1} f_i.$$

This operator η is surjective by assumption, and it is bounded, $\|\eta(\bar{f})\|_{\mathcal{H}} \leq m\|\bar{f}\|_{\bar{\mathcal{H}}}$. From the open mapping theorem, it follows that $\eta(B_1(0))$ is open, where $B_\alpha(x)$ denotes the open ball around x with radius α in the respective space. This and $\eta(0) = 0$ show that there is a number $\alpha > 0$ such that $B_{2\alpha}(0) \subset \eta(B_1(0))$. Now let $f \in \mathcal{H}$ be given. Then it holds that $\frac{\alpha}{\|f\|_{\mathcal{H}}} f \in B_{2\alpha}(0) \subset \mathcal{H}$. Hence, there exists a $y = (y_0, \ldots, y_{m-1}) \in B_1(0) \subset \bar{\mathcal{H}}$ such that $\eta(y) = \frac{\alpha}{\|f\|_{\mathcal{H}}} f$. Setting $f_i := \frac{\|f\|_{\mathcal{H}}}{\alpha} y_i \in \mathcal{H}_i$, we have $\sum_{i=0}^{m-1} f_i = f$ and

$$\|f_i\|_{\mathcal{H}} \leq \frac{\|f\|_{\mathcal{H}}}{\alpha} \|y\|_{\bar{\mathcal{H}}} \leq \frac{1}{\alpha} \|f\|_{\mathcal{H}},$$

which shows the assertion. $\square$

With the help of the above lemma, we can now see that a frame for the entire space $\mathcal{H} = \sum_{i=0}^{m-1} \mathcal{H}_i$ can be constructed by simply collecting frames for the subspaces $\mathcal{H}_i$.

Proposition 2.2.11. *Let the conditions from Lemma 2.2.10 be fulfilled and* $\mathcal{F}^{\mathcal{H}_i}$ *be frames for* $\mathcal{H}_i$, $i = 0, \ldots, m-1$. *Then, the collection*

$$\mathcal{F}^{\mathcal{H}} := \bigcup_{i=0}^{m-1} \mathcal{F}^{\mathcal{H}_i}$$

is a frame for $\mathcal{H}$.

Proof. The proof follows lines from [123, Prop. 2.7] and [38, Lemma 2.2]. Let $\mathcal{N} := \bigcup_{i=0}^{m-1} \{i\} \times \mathcal{N}_i$ denote the index set belonging to $\mathcal{F}^{\mathcal{H}}$. On the one hand, from Lemma 2.2.10, we have

$$\|f\|_{\mathcal{H}} \gtrsim \inf_{\sum_{i=0}^{m-1} f_i = f, \, f_i \in \mathcal{H}_i} \sum_{i=0}^{m-1} \|f_i\|_{\mathcal{H}}.$$

On the other hand, the converse estimate holds by the triangle inequality. Hence, from this and Lemma 2.1.6, it follows that

$$\|f\|_{\mathcal{H}}^2 \approx \inf_{\sum_{i=0}^{m-1} f_i = f, \, f_i \in \mathcal{H}_i} \sum_{i=0}^{m-1} \|f_i\|_{\mathcal{H}}^2$$

$$\approx \inf_{\sum_{i=0}^{m-1} f_i = f, \, f_i \in \mathcal{H}_i} \sum_{i=0}^{m-1} \inf_{\mathbf{c}^{(i)} \in \ell_2(\mathcal{N}_i), \, \mathbf{c}^{(i)\top} \mathcal{F}^{(i)} = f_i} \|\mathbf{c}^{(i)}\|_{\ell_2(\mathcal{N}_i)}^2$$

$$= \inf_{\mathbf{c} \in \ell_2(\mathcal{N}), \, \mathbf{c}^\top \mathcal{F} = f} \|\mathbf{c}\|_{\ell_2(\mathcal{N})}^2.$$

Another application of Lemma 2.1.6 shows the assertion. □

If we apply the above results to an overlapping domain decomposition, we directly obtain the desired result that the union of wavelet frames on the subdomains leads to an *aggregated wavelet frame* for the space $H_0^t(\Omega)$. In our construction, we will make use of Riesz bases on the subdomains. Since Riesz bases are a subclass of Hilbert frames, the theory also applies to this case.

Corollary 2.2.12. *Let $\{\Omega_i\}_{i=0}^{m-1}$ be an overlapping domain decomposition of Ω. Then, each $v \in H_0^t(\Omega)$ can be written as $v = \sum_{i=0}^{m-1} v_i$ with $v_i \in H_0^t(\Omega_i)$ and*

$$\|v_i\|_{H^t(\Omega_i)} \lesssim \|v\|_{H^t(\Omega)}.$$

Moreover, if $\Psi^{H_0^t(\Omega_i)} = \{\psi_{i,\lambda}, \, \lambda \in \Lambda_i\}$ are frames or Riesz bases for $H_0^t(\Omega_i)$, then

$$\Psi := \Psi^{H_0^t(\Omega)} := \bigcup_{i=0}^{m-1} \Psi^{H_0^t(\Omega_i)} = \{\psi_\lambda, \, \lambda \in \Lambda\}$$

is a frame for $H_0^t(\Omega)$ with index set $\Lambda := \bigcup_{i=0}^{m-1} \{i\} \times \Lambda_i$.

Proof. See Lemma 2.2.10 and Proposition 2.2.11. □

Let us now discuss a few examples and tools which will help us to construct an overlapping domain decomposition. Following [33, 104], a common approach to show that a decomposition of a domain into subdomains is overlapping in the sense of Definition 2.2.8 is by constructing a *partition of unity* with respect to the given decomposition. Such a partition of unity is defined as follows.

Definition 2.2.13. *The functions $\sigma_i : \Omega \to \mathbb{R}$, $i = 0, \dots, m-1$ are called a* partition of unity *with respect to the subdomains $\{\Omega_i\}_{i=0}^{m-1}$ of Ω, if*

(i) $\operatorname{supp} \sigma_i \subset \overline{\Omega_i}$,

(ii) $\sum_{i=0}^{m-1} \sigma_i \equiv 1$ *in* Ω,

(iii) $\sigma_i v \in H_0^t(\Omega_i)$ *for all* $v \in H_0^t(\Omega)$,

(iv) $\|\sigma_i v\|_{H^t(\Omega)} \lesssim \|v\|_{H^t(\Omega)}$.

It is obvious that the existence of a partition of unity is sufficient for the $\{\Omega_i\}_{i=0}^{m-1}$ to be an overlapping domain decomposition, since every $v \in H_0^t(\Omega)$ can be decomposed as $v = \sum_{i=0}^{m-1} \sigma_i v$. Note that a classical partition of unity, which is given by functions $v_i \in C_0^\infty(\Omega_i)$ fulfilling (i) and (ii) in Definition 2.2.13, is also a partition of unity in the above sense. This can be seen using the product rule for Sobolev functions, compare, for instance, [57, Section 5.3]. However, the conditions made here are less restrictive and can also be fulfilled in cases where such a smooth partition of unity cannot be constructed. To explain this, we present two standard examples, see, e.g., [33, 122, 123], that will also play a role in our numerical experiments.

Example 2.2.14. *The first example is the unit interval* $\Omega = (0,1)$ *with subdomains* $\Omega_0 = (0,b)$, $\Omega_1 = (a,1)$ *with* $a < b$. *This example is somewhat artificial, but sufficient to point out the basic principles. Moreover, first numerical tests will later be performed on this domain. We take* σ_0 *as a smooth function so that* $\sigma_0(x) = 1$ *for* $x \leq a$ *and* $\sigma_0(x) = 0$ *for* $x \geq b$. *A standard construction for such a function is given by*

$$\sigma_0(x) := \xi_{a,b}(x) := \frac{\omega(b-x)}{\omega(x-a) + \omega(b-x)}, \qquad (2.17)$$

where

$$\omega(x) := \begin{cases} 0, & x \leq 0 \\ \exp(-x^{-2}), & x > 0 \end{cases}.$$

Setting $\sigma_1 := 1 - \sigma_0$, *the functions* σ_0 *and* σ_1 *are smooth. Thereby, they fulfill the criteria from Definition 2.2.13.*

Example 2.2.15. *A less trivial example in two space dimensions is the L-shaped domain* (2.16). *We have seen in Subsection 2.2.4 that the construction of Riesz bases can most easily be realized on subdomains that are affine images of the unit square. Therefore, the decomposition*

$$\Omega = \Omega_0 \cup \Omega_1, \quad \Omega_0 = (-1,1) \times (-1,0), \ \Omega_1 = (-1,0) \times (-1,1) \qquad (2.18)$$

suggests itself, see also Figure 2.2. Unfortunately, for this domain decomposition, there is no continuous partition of unity. This is because every function that is equal to one on $\Omega_0 \setminus \Omega_1 = [0,1) \times (-1,0)$ *and equal to zero on* $\Omega_1 \setminus \Omega_0 = (-1,0) \times [0,1)$ *automatically has a discontinuity at the origin.*

Nevertheless, with the help of the function $\xi_{a,b}$, *a suitable non-smooth partition of unity can be constructed. Let* $(\theta(x), r(x))$ *denote the polar coordinates of a point* $x \in \mathbb{R}^2$, *taken away from the origin as sketched in Figure 2.2. Then, it is shown in [33, Section III.B] that*

$$\sigma_0 := \xi_{\frac{\pi}{2},\pi} \circ \theta, \quad \sigma_1 := 1 - \sigma_0$$

is indeed a partition of unity in our sense although the functions are not even continuous.

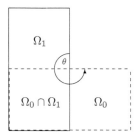

Figure 2.2: L-shaped domain with overlapping domain decomposition.

We now sketch how a partition of unity can be utilized for the construction of a dual frame. This dual frame will in general be distinct from the canonical dual frame, of which little is known and which is hard to compute in practice. For instance, the dual of a wavelet frame in $L_2(\mathbb{R})$ does not necessarily possess the wavelet structure, which means that it is not possible to only compute a single element of the dual frame and obtain the others by dilation and translation, compare [18, Section 12.1] or the general remarks in Section 2.1.2. Going to back to wavelet frames constructed by an overlapping domain decomposition, we note that although the frame $\Psi^{H_0^t(\Omega)}$ for $H_0^t(\Omega)$ itself was constructed from Riesz bases $\Psi^{(i)} := \Psi^{H_0^t(\Omega_i)}$ on the subdomains, the collection of the dual Riesz bases $\tilde{\Psi}^{(i)} := \mathbf{D}^t \tilde{\Psi}^{L_2(\Omega_i)}$ does not constitute a dual frame in general. To see this, let $\Omega = \Omega_0 \cup \Omega_1$ and $v \in H_0^t(\Omega_0 \cap \Omega_1)$, $v \neq 0$. Then, it is

$$\langle v, \tilde{\Psi}^{(0)} \rangle \Psi^{(0)} + \langle v, \tilde{\Psi}^{(1)} \rangle \Psi^{(1)} = 2v.$$

Thereby, the reconstruction formula (2.4) does not hold for the collection of the dual Riesz bases, and thus it is not a dual frame. If the dual Riesz bases, however, are weighted with the functions from the partition of unity, we indeed obtain a dual frame. This result can be formulated in terms of Gelfand frames as introduced in Subsection 2.1.3. Therefore, we start with the collection of local Riesz basis

$$\Psi^{L_2(\Omega_i)} = \mathbf{D}^t \Psi^{(i)} = \{\psi_{(i,\lambda)}, \lambda \in \Lambda_i\}$$

for $L_2(\Omega_i)$ with dual bases

$$\tilde{\Psi}^{L_2(\Omega_i)} = \mathbf{D}^{-t} \tilde{\Psi}^{(i)} = \{\tilde{\psi}_{(i,\lambda)}, \lambda \in \Lambda_i\},$$

both characterizing a range of Sobolev spaces.

Proposition 2.2.16 ([33, Prop. 3.2]). *Let and $\{\sigma_i\}_{i=0}^{m-1}$ be a partition of unity with respect to a domain decomposition $\{\Omega_i\}_{i=0}^{m-1}$ of Ω. Let*

$$\Psi^{L_2(\Omega)} := \bigcup_{i=0}^{m-1} \Psi^{L_2(\Omega_i)}, \quad \tilde{\Psi}^{L_2(\Omega)} := \bigcup_{i=0}^{m-1} \{\sigma_i \tilde{\psi}_{(i,\lambda)}, \lambda \in \Lambda_i\}.$$

Then $\Psi^{L_2(\Omega)}$ is a frame for $L_2(\Omega)$ and $\tilde{\Psi}^{L_2(\Omega)}$ is a corresponding dual frame. Moreover, $\Psi^{L_2(\Omega)}$ and $\tilde{\Psi}^{L_2(\Omega)}$ form a Gelfand frame for $(H_0^t(\Omega), L_2(\Omega), H^{-t}(\Omega))$ together with the sequence spaces $(\ell_{2,\mathbf{D}^{-t}}(\Lambda), \ell_2(\Lambda), \ell_{2,\mathbf{D}^t}(\Lambda))$, where $\Lambda = \bigcup_{i=0}^{m-1}\{i\} \times \Lambda_i$ and $\mathbf{D} = \mathrm{diag}(2^{|\lambda|})_{\lambda \in \Lambda}$.

2.2.6 Divergence-free wavelets

Divergence-free wavelet bases are very important for the numerical treatment of flow problems such as the incompressible Stokes and Navier-Stokes equation, see Section 1.4. The basic idea of the construction is to combine pairs of wavelet Riesz bases that are linked by differentiation and integration. This approach was developed in [83] for wavelets on the real line. The construction of divergence-free wavelets on the unit cube with Dirichlet boundary conditions, however, poses considerable new problems, see also the discussion in the introduction of [108]. In [109], it was possible to circumvent these difficulties and to construct divergence-free wavelets on the unit square subject to the desired boundary conditions. Although in space dimensions $d > 2$, these bases are anisotropic, the restriction to $d = 2$ yields isotropic wavelets. We now sketch the latter construction of isotropic divergence-free bases, following the outline from [38].

The starting point is a pair of biorthogonal wavelet bases

$$\Psi^{L_2(\mathcal{I})} = \{\psi_\lambda, \lambda \in \Lambda\}, \quad \tilde{\Psi}^{L_2(\mathcal{I})} = \{\tilde{\psi}_\lambda, \lambda \in \Lambda\}$$

for $L_2(\mathcal{I})$ so that the following assumptions hold.

- $\mathbf{D}^{-1}\Psi^{L_2(\mathcal{I})}$ is a Riesz basis for $H_0^1(\mathcal{I})$ and $\mathbf{D}^{-1}\tilde{\Psi}^{L_2(\mathcal{I})}$ is a Riesz basis for $H^1(\mathcal{I})$,

- $\Psi^{L_2(\mathcal{I})}$ is local in the sense that $\mathrm{diam}\,\mathrm{supp}\,\psi_\lambda \lesssim 2^{-|\lambda|}$ and that any interval of length 2^{-j} intersects the support of at most a uniformly bounded number of wavelets at level j,

$$\#\{\lambda \in \Lambda, |\lambda| = j, [x, x + 2^{-j}] \cap \mathrm{supp}\,\psi_\lambda \neq \emptyset\} \lesssim 1,$$

  and an analogous estimates holds for $\tilde{\Psi}^{L_2(\mathcal{I})}$,

- and there is a $\lambda^0 \in \Lambda$ with $|\lambda^0| = j_0$ such that $\tilde{\psi}_\lambda$ is constant.

The first conditions are standard assumptions, whereas the third assumption can be fulfilled by a basis transformation if the dual wavelets on the level j_0 of the original basis can reproduce constants. This is true for standard spline bases. In addition, it is assumed that the corresponding generators

$$\Phi_j = \{\phi_{j,k}, 1 \leq k \leq N_j\}, \quad \tilde{\Phi}_j = \{\tilde{\phi}_{j,k}, 1 \leq k \leq N_j\}$$

are biorthogonal, uniformly stable and local in the above sense with constants independent of $j \geq j_0$. Moreover, it is assumed that for each j, $\int_{\mathcal{I}} \phi_{j,k}(x)\,\mathrm{d}x$ is independent of k and that the generators are ordered in the sense that $\inf \mathrm{supp}\,\phi_{j,k} \leq \inf \mathrm{supp}\,\phi_{j,k'}$ if $k \leq k'$.

With such a pair of biorthogonal wavelet bases at hand, a new pair of biorthogonal wavelet systems is constructed, that is linked to the original bases by differentiation and integration. To do so, with $\Lambda^0 := \Lambda \setminus \lambda^0$ set

$$\Psi^+ := \{\psi_\lambda^+, \, \lambda \in \Lambda^0\}, \quad \psi_\lambda^+(x) := \int_0^x 2^{|\lambda|}\psi_\lambda(y)\,\mathrm{d}y$$

and

$$\tilde{\tilde{\Psi}} := \{\tilde{\tilde{\psi}}_\lambda, \, \lambda \in \Lambda^0\}, \quad \tilde{\tilde{\psi}}_\lambda(x) := -2^{-|\lambda|}\tilde{\psi}'_\lambda(x).$$

In [109, Th. 4.2] and [38, Prop. 2.5], it is shown that Ψ^+ and $\tilde{\tilde{\Psi}}$ are biorthogonal systems in $L_2(\mathcal{I})$ and local, Ψ^+ is a Riesz basis for $L_2(\mathcal{I})$ and $\mathbf{D}^{-2}\Psi^+$ is a Riesz basis for $H_0^2(\mathcal{I})$. Moreover, we transform the generator functions by

$$\phi_{j,k}^+(x) := \int_0^x 2^{j+1}(\phi_{j,k+1}(y) - \phi_{j,k}(y))\,\mathrm{d}y, \quad \tilde{\tilde{\phi}}_{j,k}(x) = -2^{-(j+1)}\sum_{p=k+1}^{N_j}\tilde{\phi}'_{j,p}(x).$$

Writing $\Psi_{[j]}^+ = \{\psi_\lambda^+, |\lambda| = j\}$ and defining $\Phi_{[j]}$, $\Psi_{[j]}$ analogously, it is shown in [38, Section 2.4] and [109, Section 5] that from these sets, via tensorization we can construct a Riesz basis for $H_0^2(\mathcal{I})$ in the form

$$\bigcup_{j \geq j_0} 2^{-j}\left(2^{-(j+1)}\Psi_{[j+1]}^+ \otimes \Phi_{[j]}^+ \cup \Phi_{[j]}^+ \otimes \Psi_{[j+1]}^+ \cup \Psi_{[j]}^+ \otimes \Psi_{[j]}^+\right). \qquad (2.19)$$

The last fact is the key to employing a result from [65], which says that the curl-operator

$$\operatorname{curl}(v) := \left(\frac{\partial v}{\partial x_2}, -\frac{\partial v}{\partial x_1}\right)^\top$$

is an isomorphism from $H_0^2(\mathcal{I}^2)$ to

$$V(\mathcal{I}^2) := \{v \in (H_0^1(\mathcal{I}^2))^2, \, \operatorname{div} v = 0\}.$$

Thereby, the design principle is to apply the negative curl-operator to the basis constructed in (2.19). Then, the desired divergence-free wavelet basis is directly obtained. Moreover, we can also give an explicit formula for the elements of the dual basis.

Proposition 2.2.17 ([109, Prop. 5.5] and [38, Prop. 2.8]). *Let* $\Lambda_j^0 := \{\lambda \in \Lambda^0, \, |\lambda| = j\}$. *The collection*

$$\Psi^{V(\mathcal{I}^2)} := \bigcup_{j \geq j_0} 2^{-j}(\{[-\psi_\lambda^+ \otimes (\phi_{j,k+1} - \phi_{j,k}), \psi_\lambda \otimes \phi_{j,k}^+]^\top, \lambda \in \Lambda_{j+1}^0, 1 \leq k \leq N_j - 1\}$$

$$\cup\{[-\phi_{j,k}^+ \otimes \psi_\lambda, (\phi_{j,k+1} - \phi_{j,k}) \otimes \psi_\lambda^+]^\top, \lambda \in \Lambda_{j+1}^0, 1 \leq k \leq N_j\}$$

$$\cup\{[-\psi_\lambda^+ \otimes \psi_\mu, \psi_\lambda \otimes \psi_\mu^+]^\top, \lambda, \mu \in \Lambda_j^0\})$$

is a Riesz basis for $V(\mathcal{I}^2)$ with the dual basis

$$\tilde{\Psi}^{V(\mathcal{I}^2)} := \bigcup_{j \geq j_0} 2^j (\{[0, \tilde{\psi}_\lambda \otimes \bar{\tilde{\phi}}_{j,k}]^\top, \; \lambda \in \Lambda_{j+1}^0, \; 1 \leq k \leq N_j\}$$

$$\cup \{[-\bar{\tilde{\phi}}_{j,k} \otimes \tilde{\psi}_\lambda, 0]^\top, \; \lambda \in \Lambda_{j+1}^0, \; 1 \leq k \leq N_j\}$$

$$\cup \{[-\bar{\tilde{\psi}}_\lambda \otimes \tilde{\psi}_\mu, \tilde{\psi}_\lambda \otimes \bar{\tilde{\psi}}_\mu], \; \lambda, \mu \in \Lambda_j^0\}).$$

Both the primal and dual basis are local in the above sense.

In a similar fashion as with the standard wavelet construction, these bases can be generalized to bases on affine images of the unit square, $\Omega = \kappa(\mathcal{I}^2)$. To make sure that the wavelets remain divergence-free, the transformation is slightly modified to

$$\psi_\lambda^{V(\Omega)}(x) = \frac{D\kappa(\kappa^{-1}(x))\psi_\lambda^{V(\mathcal{I}^2)}(\kappa^{-1}(x))}{|\det D\kappa(\kappa^{-1}(x))|^{1/2}},$$

where $\psi_\lambda^{V(\mathcal{I}^2)}$ are the wavelets in $V(\mathcal{I}^2)$, see [17, Section 5]. The dual basis can be transformed in the same way.

So far, we have sketched the construction of divergence-free wavelet bases on affine images of the unit cube. To construct divergence-free wavelet frames by an overlapping domain decomposition as in Subsection 2.2.5, it is important to know whether the collection of divergence-free wavelet bases on the subdomains constitutes a frame on the entire domain. It turns out that this is true under the same conditions as in Subsection 2.2.5. To show this, we will employ the following lemma.

Lemma 2.2.18 ([86, Lemma 2]). *Let Ω_i, $i = 0, \ldots, m-1$ be subdomains of Ω such that $H_0^1(\Omega)^2 = H_0^1(\Omega_0)^2 + \ldots + H_0^1(\Omega_{m-1})^2$. Then,*

$$V(\Omega) = V(\Omega_0) + \ldots + V(\Omega_{m-1}).$$

With the help of this lemma, the frame property follows quite easily.

Proposition 2.2.19 ([38, Lemma 2.2]). *Assume that $\{\Omega_i\}_{i=0}^{m-1}$ is an overlapping domain decomposition of Ω in the sense of Definition 2.2.8 with $t = 1$. Let $\Psi^{V(\Omega_i)}$ be frames or Riesz bases for $V(\Omega_i)$ equipped with the Sobolev norm. Then, their union $\Psi^{V(\Omega)} := \bigcup_{i=0}^{m-1} \Psi^{V(\Omega_i)}$ is a frame for $V(\Omega)$.*

Proof. Let us recall the short argument from [38]. Because the subdomains are overlapping, we have $H_0^1(\Omega)^2 = H_0^1(\Omega_0)^2 + \ldots + H_0^1(\Omega_{m-1})^2$. By Lemma 2.2.18, it is $V(\Omega) = V(\Omega_0) + \ldots + V(\Omega_{m-1})$. These spaces are closed with respect to the Sobolev norm. Hence, the assertion follows from Proposition 2.2.11. $\square$

The construction of divergence-free wavelet frames can also be embedded into the concept of Gelfand frames. For this, we refer to [17, Section 5]. There, however, the conditions on the overlap of the subdomains are more restrictive than in Proposition 2.2.19. In particular, the standard example of the L-shaped domain is excluded.

2.3 Discretization with wavelet frames

In the following section, we explain how equations of the type

$$\mathcal{L}u + \mathcal{G}(u) = f, \tag{2.20}$$

as formulated in Section 1.2, or the weak form of the stationary Navier-Stokes equation, respectively, can be discretized with the help of a wavelet frame. In our case, discretization means that the equation is reformulated as an equation in $\ell_2(\Lambda)$, in other words, as an infinite-dimensional system of scalar equations. The outline partly follows standard lines such as [104]. We describe the discretization of classical semilinear equations and then point out how to adapt the concept to the stationary Navier-Stokes equation.

2.3.1 Semilinear elliptic equations

In this subsection, let $\Psi := \Psi^{H_0^t(\Omega)} = \{\psi_\lambda, \lambda \in \Lambda\}$ be a wavelet frame for $H_0^t(\Omega)$ constructed from local Riesz bases $\Psi^{(i)} := \Psi^{H_0^t(\Omega_i)}$ by means of an overlapping domain decomposition as sketched in the previous section. Recall from Section 2.1.2 that, corresponding to such a frame, we have an injective analysis operator $F : H_0^t(\Omega) \to \ell_2(\Lambda)$. Hence, we can multiply Equation (2.20) from the left with F. This gives the equivalent formulation

$$F\mathcal{L}u + F\mathcal{G}(u) = Ff.$$

Moreover, we have at hand a surjective synthesis operator $F^* : \ell_2(\Lambda) \to H_0^t(\Omega)$. Hence, there exists a $\mathbf{u} \in \ell_2(\Lambda)$ with $u = F^*\mathbf{u}$. Inserting this into the equation gives a new equation in $\ell_2(\Lambda)$, namely

$$F\mathcal{L}F^*\mathbf{u} + F\mathcal{G}(F^*\mathbf{u}) = Ff.$$

In shorter form, this equation can be written as

$$\mathbf{A}\mathbf{u} + \mathbf{G}(\mathbf{u}) = \mathbf{f}, \tag{2.21}$$

where $\mathbf{A} := F\mathcal{L}F^* : \ell_2(\Lambda) \to \ell_2(\Lambda)$ represents the discretized version of the linear part, $\mathbf{G} := F\mathcal{G}F^* : \ell_2(\Lambda) \to \ell_2(\Lambda)$ is the discrete equivalent of the nonlinearity and

$$\mathbf{f} := Ff = ((\langle f, \psi_\lambda \rangle)_{\lambda \in \Lambda} \in \ell_2(\Lambda).$$

is the discrete right-hand side. In the following, we study the terms on the left-hand side of this equation in a little more detail, starting with the linear part $\mathbf{A}$. Note first that the application of this mapping to a vector $\mathbf{v} = (v_\lambda)_{\lambda \in \Lambda} \in \ell_2(\Lambda)$ can be written as

$$\mathbf{A}\mathbf{v} = F\mathcal{L}F^*\mathbf{v} = \left(\sum_{\lambda \in \Lambda} \langle \mathcal{L}\psi_\lambda, \psi_\mu \rangle v_\lambda \right)_{\mu \in \Lambda},$$

which is equivalent to the multiplication with the symmetric matrix

$$(\langle \mathcal{L}\psi_\lambda, \psi_\mu \rangle)_{\mu,\lambda \in \Lambda} = (a(\psi_\lambda, \psi_\mu))_{\mu,\lambda \in \Lambda}.$$

Therefore, with a slight abuse of notation, the mapping $\mathbf{A}$ will be identified with this corresponding matrix $(a(\psi_\lambda, \psi_\mu))_{\mu,\lambda \in \Lambda}$ and be called *stiffness matrix*.

Remark 2.3.1. *In the literature, the wavelets ψ_λ are often considered as elements from a basis or frame in $L_2(\Omega)$ such that the rescaled version $\Psi^{H_0^t(\Omega)} = \mathbf{D}^{-t}\Psi^{L_2(\Omega)}$ is a basis or frame for $H_0^t(\Omega)$. With this alternative notation, the discretized linear part has the form $\mathbf{A} = \mathbf{D}^{-t}F\mathcal{L}F^*\mathbf{D}^{-t}$ and the matrix coefficients are given by $2^{-t(|\mu|+|\lambda|)}a(\psi_\lambda, \psi_\mu)$. However, with the scaling already incorporated into the basis, this is the same matrix as defined above.*

Some important properties of the stiffness matrix are summarized in the following lemma.

Lemma 2.3.2 ([34, Section 4], [104, Section 2.2]). *The mapping $\mathbf{A}$ is a bounded, symmetric positive semi-definite operator from $\ell_2(\Lambda)$ to $\ell_2(\Lambda)$. It holds that $\operatorname{ran}\mathbf{A} = \operatorname{ran}F$ and $\ker \mathbf{A} = \ker F^*$. Moreover, $\mathbf{A}$ is boundedly invertible on its range.*

Proof. The mapping $\mathbf{A}$ is bounded, because it is a composition of bounded operators. The symmetry of $a(\cdot, \cdot)$ carries over to $\mathbf{A}$. Moreover, for $\mathbf{v} \in \ell_2(\Lambda)$, we have

$$\langle \mathbf{A}\mathbf{v}, \mathbf{v} \rangle_{\ell_2(\Lambda)} = \langle F\mathcal{L}F^*\mathbf{v}, \mathbf{v} \rangle_{\ell_2(\Lambda)} = \langle \mathcal{L}(F^*\mathbf{v}), F^*\mathbf{v} \rangle = a(F^*\mathbf{v}, F^*\mathbf{v}). \qquad (2.22)$$

Since $a(\cdot, \cdot)$ is assumed to be elliptic, it follows that $\mathbf{A}$ is symmetric positive semi-definite. Because $\mathcal{L}$ and F^* are onto, it is $\operatorname{ran}\mathbf{A} = \operatorname{ran}F$, whereas $\ker \mathbf{A} = \ker F^*$ follows from F and $\mathcal{L}$ being injective. Let $\mathbf{B} := FS^{-1}\mathcal{L}^{-1}S^{-1}F^* : \ell_2(\Lambda) \to \ell_2(\Lambda)$. This is again a composition of bounded operators, thus bounded. Because of $S = F^*F$, we have

$$\mathbf{A}\mathbf{B} = F\mathcal{L}^{-1}F^*FS^{-1}\mathcal{L}S^{-1}F^* = FS^{-1}F^*$$

and, similarly, $\mathbf{B}\mathbf{A} = FS^{-1}F^*$. Since $FS^{-1}F^* = \mathbf{Q}$ is the orthogonal projector on $\operatorname{ran}F = \operatorname{ran}\mathbf{A}$, see Subsection 2.1.2, it is $\mathbf{B} = \mathbf{A}^{-1}$ on $\operatorname{ran}\mathbf{A}$, which shows the assertion. $\qquad \square$

In domain decomposition methods, the structure of the frame as a collection of Riesz bases will be exploited. To this end, it will be useful to consider the blocks

$$\mathbf{A}^{(i,j)} := (a(\psi_\lambda, \psi_\mu))_{\mu \in \Lambda_i, \lambda \in \Lambda_j} \qquad (2.23)$$

of the matrix $\mathbf{A}$ corresponding to the wavelets from the Riesz bases on the subdomains Ω_i and Ω_j. In particular, the properties of the diagonal blocks $\mathbf{A}^{(i,i)}$ will be of special importance. These properties will be summarized in the following lemma, which in similar form has been shown in [25].

Lemma 2.3.3. *The diagonal blocks $\mathbf{A}^{(i,i)}$ of $\mathbf{A}$ are symmetric positive definite and boundedly invertible on $\ell_2(\Lambda_i)$. Moreover,*

$$\|\mathbf{v}\|_{\mathbf{A}^{(i,i)}} := \left(\langle \mathbf{A}^{(i,i)}\mathbf{v}, \mathbf{v} \rangle_{\ell_2(\Lambda_i)} \right)^{1/2}$$

is a norm on $\ell_2(\Lambda_i)$ equivalent to $\| \cdot \|_{\ell_2(\Lambda_i)}$.

Proof. Because $\Psi^{(i)}$ is a Riesz basis, the analysis operator corresponding to $\Psi^{(i)}$ is surjective, hence ran $\mathbf{A}^{(i,i)} = \ell_2(\Lambda_i)$, therefore the matrix is boundedly invertible, see Lemma 2.3.2. Moreover, as in the proof above, $\langle \mathbf{A}^{(i,i)}\mathbf{v}, \mathbf{v} \rangle_{\ell_2(\Lambda_i)} = a(\mathbf{v}^\top \Psi^{(i)}, \mathbf{v}^\top \Psi^{(i)})$. Because $a(\cdot, \cdot)$ is bounded and elliptic, and $\Psi^{(i)}$ is a Riesz basis, it is $a(\mathbf{v}^\top \Psi^{(i)}, \mathbf{v}^\top \Psi^{(i)}) \eqsim \|\mathbf{v}^\top \Psi^{(i)}\|^2_{H^t(\Omega_i)} \eqsim \|\mathbf{v}\|^2_{\ell_2(\Lambda_i)}$. $\qquad\square$

Let us now turn to the discrete counterpart $\mathbf{G}$ of the nonlinear term $\mathcal{G}$. The application of this mapping to a vector $v \in \ell_2(\Lambda)$ can be written as

$$\mathbf{G}(\mathbf{v}) = (\langle \mathcal{G}(\mathbf{v}^\top \Psi), \psi_\lambda \rangle)_{\lambda \in \Lambda}.$$

This discretized nonlinearity inherits important properties of $\mathcal{G}$. For instance, if G is monotone, and thus $\mathcal{G}$, this also holds for $\mathbf{G}$,

$$\langle \mathbf{G}(\mathbf{v}) - \mathbf{G}(\mathbf{w}), \mathbf{v} - \mathbf{w} \rangle_{\ell_2(\Lambda)} \geq 0, \quad \mathbf{v}, \mathbf{w} \in \ell_2(\Lambda). \tag{2.24}$$

The stability relation (1.15) carries over to $\mathbf{G}$ in the sense that

$$\|\mathbf{G}(\mathbf{v}) - \mathbf{G}(\mathbf{w})\|_{\ell_2(\Lambda)} \leq \hat{C}_G(\max\{\|\mathbf{Q}\mathbf{v}\|_{\ell_2(\Lambda)}, \|\mathbf{Q}\mathbf{w}\|_{\ell_2(\Lambda)}\})\|\mathbf{v} - \mathbf{w}\|_{\ell_2(\Lambda)}, \tag{2.25}$$

where $\mathbf{Q}$ is the orthogonal projector on ran F and $\hat{C}_G(x) = \|F\| \cdot \|F^*\| \cdot C_G(\|F^*\|x)$. The derivation of the above estimates can be found in [76, Lemma 2.35].

It remains to explain how the solutions of the continuous and the discretized version are related, in particular considering possible overcompleteness of the system Ψ. This will be targeted by the following lemma, which in similar form has been shown in [27, 75].

Lemma 2.3.4. *Assume that Equation (2.20) has a unique solution $u \in H_0^t(\Omega)$. Then, $\bar{\mathbf{u}} := FS^{-1}u$ solves (2.21). The set of all solutions of (2.21) can be written as $\mathbf{u} + \ker F^*$, where $\mathbf{u}$ is any such solution, for instance $\mathbf{u} = \bar{\mathbf{u}}$. From any $\mathbf{u} \in \ell_2(\Lambda)$ fulfilling Equation (2.21), the solution of (2.20) can be retrieved by $u = F^*\mathbf{u} = \mathbf{u}^\top \Psi$.*

Proof. A simple calculation shows

$$\mathbf{A}\bar{\mathbf{u}} + \mathbf{G}(\bar{\mathbf{u}}) = F\mathcal{L}F^*FS^{-1}u + F\mathcal{G}(F^*FS^{-1}u) = F(\mathcal{L}u + \mathcal{G}(u)) = Ff = \mathbf{f},$$

hence $\bar{\mathbf{u}}$ is indeed a discrete solution. Because F is injective, Equation (2.20) is equivalent to $\mathcal{L}F^*\mathbf{u} + \mathcal{G}(F^*\mathbf{u}) = f$, which, by assumption, holds if and only if $F^*\mathbf{u} = u$. This shows the characterization of the solution set. If, in turn, $\mathbf{u} \in \ell_2(\Lambda)$ is any discrete solution, by the above it is $\mathbf{u} - \bar{\mathbf{u}} \in \ker F^*$, hence $F^*\mathbf{u} = F^*\bar{\mathbf{u}} = u$. $\qquad\square$

2.3.2 Stationary Navier-Stokes equation

The discretization of the LeRay weak form (1.25) of the stationary, incompressible Navier-Stokes equation follows similar lines as for the semilinear equations from the previous subsection. For the Navier-Stokes equation, we make use of a divergence-free wavelet frame $\Psi := \Psi^{V(\Omega)} = \{\psi_\lambda, \lambda \in \Lambda\}$ for V, which is a collection of local divergence-free wavelet Riesz bases $\Psi^{(i)} := \Psi^{V(\Omega_i)}$. The construction of such systems has been outlined in Subsection 2.2.6. Moreover, let F^* and F be the synthesis and analysis operators corresponding to this frame.

Let us first note that, under the conditions from Lemma 1.4.1, the weak form (1.25) of the Navier-Stokes equation can be written as

$$\mathcal{L}u + \operatorname{Re}\mathcal{G}(u) = \operatorname{Re} f, \tag{2.26}$$

with $\mathcal{L}v(w) = a(v, w)$ and $\mathcal{G}(v) = (v \cdot \nabla)v$. As in the previous subsection, by multiplying this equation with F from the left and writing $u = F^*\mathbf{u}$, $\mathbf{u} \in \ell_2(\Lambda)$, we end up with a discretized equation

$$\mathbf{Au} + \operatorname{Re}\mathbf{G}(\mathbf{u}) = \operatorname{Re}\mathbf{f} \tag{2.27}$$

in $\ell_2(\Lambda)$, where $\mathbf{A} = F\mathcal{L}F^*$, $\mathbf{G} = F\mathcal{G}F^*$ and $\mathbf{f} = Ff$. Let us take a look at the terms on the left-hand side. As a composition of bounded linear mappings, the operator $\mathbf{A}$ is bounded operator on $\ell_2(\Lambda)$. As before, we can identify this operator with the corresponding stiffness matrix $\mathbf{A} = (a(\psi_\mu, \psi_\lambda))_{\mu,\lambda \in \Lambda}$. Because $a(\cdot, \cdot)$ is bounded and elliptic on $V \times V$, the same arguments as in Lemma 2.3.2, with $H_0^t(\Omega)$ replaced by V, show that $\mathbf{A}$ is boundedly invertible on $\operatorname{ran}\mathbf{A} = \operatorname{ran} F$ and that $\ker\mathbf{A} = \ker F^*$. Moreover, with the arguments from Lemma 2.3.3, we see that its diagonal blocks $\mathbf{A}^{(i,i)}$ defined analogously to (2.23) are boundedly invertible.

Concerning the discretized nonlinear part, from Lemma 1.4.1, we see that $\mathcal{G}$ indeed maps V into V' and that $\mathcal{G}$ is stable in the sense that $\|\mathcal{G}(v)\|_{V'} \lesssim \|v\|_V^2$. From the mapping properties of F and F^*, it follows that $\mathbf{G}$ is well-defined as a mapping from $\ell_2(\Lambda)$ to $\ell_2(\Lambda)$. Moreover, the application of this mapping to a vector $\mathbf{v} \in \ell_2(\Lambda)$ can now be written as

$$\mathbf{G}(\mathbf{v}) = \left(\int_\Omega \psi_\lambda \cdot (v \cdot \nabla)v \, dx \right)_{\lambda \in \Lambda}, \quad v = \mathbf{v}^\top \Psi.$$

To understand the connection between (1.25) and (2.27), recall first that from Theorem 1.4.2, we know that Equation (1.25) has a unique solution $u \in V$ if the Reynolds number and the norm of the right-hand side are sufficiently small. Under these conditions, the same arguments as in Lemma 2.3.4 show that the solutions of the discretized Navier-Stokes equation (2.27) are given by $\mathbf{u} + \ker F^*$, where $\mathbf{u}$ is a particular solution such as $\mathbf{u} = FS^{-1}u$.

Chapter 3

Nonlinear wavelet approximation and regularity

To analyze the efficiency of numerical methods for operator equations, concepts and tools from approximation theory play a vital role. A core point of interest of this theory is the approximation of a given element from a Banach space, or even a Hilbert space, by approximations from nested subsets. Although in numerical analysis, the object to approximate is typically given only implicitly as the solution of a given problem, much theoretical insight can however we gained by taking the point of view of approximation theory. This is because we will see that standard uniform methods for the solution of partial differential equations can be described in terms of approximation from linear spaces, whereas adaptive methods can be compared to approximation from nonlinear spaces. These two types of approximation strategies are called *linear* and *nonlinear approximation*.

To evaluate the efficiency of an approximation scheme, we have to understand how fast the approximations converge to the element we approximate and measure the rate in terms of the effort needed for the computation of the approximations. A good proxy for this effort is typically given by the dimension or degrees of freedom of the subsets from which we calculate the approximation. It turns out that, when approximating a function v, in many cases including approximation with wavelets, this approximation rate can be characterized in terms of the smoothness of v in classical smoothness spaces. In particular, linear approximation is related with Sobolev smoothness, whereas nonlinear approximation corresponds to regularity measured in a different scale, a particular scale of *Besov spaces*. Therefore, we introduce these spaces and recall some important regularity results with respect to this scale. This will help us to understand if and for which cases we can expect a benefit from applying adaptive methods.

When dealing with nonlinear problems, as we do in this thesis, we will see later that we have to confine ourselves to approximations with wavelets from *tree-structured* index sets. Hence, we introduce this concept and explain how the additional restriction to such index sets affects the approximation rate. Fortunately, it turns out that the impact is very little.

3.1 Basics of approximation theory

We now outline some very basic concepts from approximation theory, roughly following the standard references [23, 52, 53]. For a general introduction to approximation theory, we also refer to [19, 93]. Proofs to the statements given here can, for instance, be found in [53, Ch. 7]. In a first step, we introduce *approximation spaces*. In the case of nonlinear wavelet approximation, these spaces are linked to classical Besov spaces, which will also be introduced in this section.

3.1.1 Approximation spaces

We consider a normed linear space X, typically a Banach space, and nested subsets X_n, $n \in \mathbb{N}$, that are asymptotically dense in X, i.e., $\mathrm{clos}_X \bigcup_{n \in \mathbb{N}} X_n = X$. We define the *approximation error* as

$$E_n(v) := \mathrm{dist}(v, X_n) := \inf_{g \in X_n} \|v - g\|_X.$$

Because the spaces X_n are nested and asymptotically dense, $E_n(v)$ converges to zero when n goes to infinity. The rate of convergence, however, in general depends on the properties of v. To collect all elements in X for which the approximation error has a similar decay, we define the *approximation spaces*

$$A_q^s(X) := \{v \in X, \ |v|_{A_q^s(X)} := \|(2^{sn} E_n(v))_{n \in \mathbb{N}}\|_{\ell_q(\mathbb{N})} < \infty\}$$

with parameters $s > 0$ and $0 \leq q \leq \infty$. The quantity

$$\|v\|_{A_q^s(X)} := \|v\|_X + |v|_{A_q^s(X)}$$

is a norm on $A_q^s(X)$ for $q \geq 1$ and a quasi-norm for $0 < q < 1$. These spaces are nested in the sense that $A_{q_1}^{s_1}(X) \subset A_{q_2}^{s_2}(X)$ whenever $s_1 > s_2$. If $s_1 = s_2$, the embedding still holds when $q_1 \leq q_2$, , see also [52, Section 4.1]. For $q = \infty$, v belongs to the space $A_\infty^s(X)$ if and only if $E_n(v) \lesssim 2^{-sn}$, which means that the approximation error decays with rate 2^{-s} in n.

When approximating the solution of a boundary value problem, the space X typically is the Sobolev space in which the problem is formulated. The sets X_n are linear or nonlinear subspaces from which the approximations are calculated. In this setting, our aim will be to find norm equivalences between suitable smoothness spaces and the above approximation spaces. This will allow us to describe the approximate rate in terms of the smoothness of the function we want to approximate. One important type of such smoothness spaces will be introduced in the following.

3.1.2 Besov spaces and their interpolation

In this subsection, following [23, 52], we give the definition and some important facts on Besov spaces. For further reading on these spaces and proofs to the facts given here, we also refer to [53, 116, 117, 118]. At first, for a bounded Lipschitz domain Ω, a function $v : \Omega \to \mathbb{R}$ and $h > 0$, we define the *difference operator* Δ_h by

$$\Delta_h v := v(\cdot + h) - v(\cdot).$$

Recursively, the n-th iterated difference operator is then given by $\Delta_h^n v := \Delta_h(\Delta_h^{n-1} v)$ with $\Delta_h^0 v := v$. To make sure that this expression is well-defined, i.e., that the points in which v is evaluated remain within Ω, the domain of the n-th iterated difference $\Delta_h^n v$ is restricted to

$$\Omega_{h,n} := \{x \in \Omega,\ x + kh \in \Omega,\ 0 \leq k \leq n\}.$$

For $v \in L_p(\Omega)$, the n-th modulus of smoothness is now defined as

$$\omega_n(v,t)_p := \sup_{|h| \leq t} \|\Delta_h^n f\|_{L_p(\Omega_{h,n})}.$$

Clearly, $\omega_n(v,t)_p$ is monotonically increasing in t. With $\alpha > 0$ and n being the smallest integer larger than α, the *Besov space* $B_{p,q}^\alpha(\Omega)$, $0 < p, q \leq \infty$ is now defined as the set of all functions $v \in L_p(\Omega)$, for which

$$|v|_{B_{p,q}^\alpha(\Omega)} := \begin{cases} \left(\int_0^\infty \left(t^{-\alpha} \omega_n(v,t)_p \right)^q \frac{dt}{t} \right)^{1/q}, & q \in (0,\infty) \\ \sup_{t>0} t^{-\alpha} \omega_n(v,t)_p, & q = \infty \end{cases}$$

is finite. For $1 \leq p, q \leq \infty$, the expression

$$\|v\|_{B_{p,q}^\alpha(\Omega)} := \|v\|_{L_p(\Omega)} + |v|_{B_{p,q}^\alpha(\Omega)}$$

is a norm on $B_{p,q}^\alpha(\Omega)$, and still a quasi-norm if only $0 < p, q \leq \infty$. Let us make a few remarks on the roles of the parameters α, p and q. The parameter α is typically interpreted as the order of smoothness, whereas p stands for the L_p-norm in which smoothness is measured. The parameter q is a secondary parameter in the sense that for any $q_1, q_2 > 0$ and $\alpha_1 > \alpha_2$, it is

$$B_{p,q_1}^{\alpha_1}(\Omega) \subset B_{p,q_2}^{\alpha_2}(\Omega). \tag{3.1}$$

Moreover, it is $B_{p,q_1}^\alpha(\Omega) \subset B_{p,q_2}^\alpha(\Omega)$ if $q_1 \leq q_2$ and $B_{p_1,q}^\alpha(\Omega) \subset B_{p_2,q}^\alpha(\Omega)$ if $p_1 \geq p_2$. The latter embeddings are due to embeddings of the L_p-spaces on bounded domains, see [116] for detailed proofs. A more difficult question, however, is how Besov spaces with different parameters α and p are embedded into each other, when both of them are varied. This relation is clarified by the following result, see also [116]. In particular, this corresponds to the embedding of Sobolev spaces from Theorem 1.1.2.

Proposition 3.1.1 ([23, Corollary 3.7.1]). *Let $\alpha_1 > \alpha_2 > 0$ and $\infty \geq p_2 > p_1 > 0$ with $\alpha_1 - \alpha_2 = d(\frac{1}{p_1} - \frac{1}{p_2})$. Then we have the continuous embedding*

$$B_{p_1,q}^{\alpha_1}(\Omega) \subset B_{p_2,q}^{\alpha_2}(\Omega) \tag{3.2}$$

for all $q > 0$.

Moreover, many Besov spaces with parameters $p = q$ are Sobolev spaces. If α is not an integer, it is $B_{p,p}^{\alpha}(\Omega) = W^{\alpha}(L_p(\Omega))$. For the special case $p = 2$, this equality holds for all α, hence

$$B_{2,2}^{\alpha}(\Omega) = H^{\alpha}(\Omega),$$

see, e.g., [52, Section 4.5]. In many applications, we have information about a function being contained in two different Besov spaces. Then, we would like to conclude that it also belongs to all Besov spaces that, in a way, lie in between these two spaces. To describe this precisely, let us make some remarks on *interpolation spaces*, following [8, 23, 54]. Let X_0, X_1 be Banach spaces that are continuously embedded in a Hausdorff topological vector space to ensure that their intersection and sum are well-defined. An interpolation space is a space X so that $X_0 \cap X_1 \subset X \subset X_0 + X_1$ and that any linear operator that is bounded from X_0 to itself and from X_1 to itself is also bounded as an operator from X to X. One way to construct such spaces is by a *Peetre K-functional*. For $f \in X_0 + X_1$ and $t > 0$, this functional is defined as

$$K(f,t) := \inf\{\|f_0\|_{X_0} + t\|f_1\|_{X_1}, \ f = f_0 + f_1, \ f_0 \in X_0, \ f_1 \in X_1\}.$$

With the help of this functional, and for $0 < \theta < 1$, $0 < q \leq \infty$, an interpolation space $(X_0, X_1)_{\theta,q}$ can be defined as the set of all $f \in X_0 + X_1$, for which

$$|f|_{(X_0,X_1)_{\theta,q}} := \begin{cases} \left(\int_0^{\infty} [t^{-\theta}K(f,t)]^q \frac{dt}{t}\right)^{1/q}, & 0 < q < \infty \\ \sup_{t>0} t^{-\theta} K(f,t), & q = \infty \end{cases}$$

is finite. In the case of Besov spaces, these interpolation spaces can explicitly be characterized as Besov spaces with intermediate parameters.

Proposition 3.1.2 ([54, Corollary 6.3]). *Let $\alpha_0, \alpha_1 > 0$, $0 < p_0, p_1, q_0, q_1 \leq \infty$ and $0 < \theta < 1$. Then, holds that*

$$(B_{p_0,q_0}^{\alpha_0}(\Omega), B_{p_1,q_1}^{\alpha_1}(\Omega))_{\theta,q} = B_{p,q}^{\alpha}(\Omega), \tag{3.3}$$

where $\alpha = \theta\alpha_0 + (1-\theta)\alpha_1$, $\frac{1}{p} = \frac{\theta}{p_0} + \frac{1-\theta}{p_1}$ and $\frac{1}{q} = \frac{\theta}{q_0} + \frac{1-\theta}{q_1}$.

The embedding (3.2) of Besov spaces into other Besov spaces and the embedding into Sobolev spaces are often visualized with the help of a *DeVore-Triebel diagram*. These diagrams will later be helpful to better understand some of the proofs in which such embeddings are used. A first example is shown in Figure 3.1. There, a point in the $(\frac{1}{\tau}, \alpha)$-plane represents the Besov space $B_{\tau,\tau}^{\alpha}(\Omega)$. The embedding (3.1) tells us that such a space is contained in all spaces corresponding to points below or at the lower right of the given point. Let us start from a point $(\frac{1}{\tau_1}, \alpha_1)$ that lies on the line $\frac{1}{\tau} = \frac{\alpha}{d} + \frac{1}{2}$. By the relation (3.2), the corresponding Besov space $B_{\tau_1,\tau_1}^{\alpha_1}(\Omega)$ is embedded in $L_2(\Omega)$. This also holds for all spaces above the line, even for other secondary parameters. Therefore, this line is called L_2-embedding line. In a similar fashion, we can visualize the embedding into $H^t(\Omega) = B_{2,2}^t(\Omega)$ as sketched in the figure.

Proposition 3.1.2 can be interpreted in the sense that if a function is contained in two spaces represented by points in the DeVore-Triebel diagram, then it is also contained in all the spaces on the line segment connecting these points.

So far, we have only introduced Besov spaces *without* boundary conditions. In our applications, however, we will mostly be concerned with problems involving Dirichlet boundary conditions. To define Besov spaces with such boundary conditions, we first denote by

$$C^\infty(\partial\Omega, k) := \{v \in C^\infty(\Omega), \ |v(x)| \lesssim (\text{dist}(x, \partial\Omega))^{k+1}\}$$

the space of all smooth functions on Ω that vanish with order k on the boundary. Then, the Besov space with homogeneous boundary conditions is simply defined as

$$B_{p,q}^\alpha(\partial\Omega, k) := \text{clos}_{B_{p,q}^\alpha(\Omega)} C^\infty(\partial\Omega, k-1).$$

From time to time, we will also need Besov spaces with boundary conditions prescribed by inner boundaries that are induced by an overlapping domain decomposition $\{\Omega_i\}_{i=0}^{m-1}$. Hence, with l being be the order of polynomial exactness of the primal wavelet basis, we denote by $B_{p,q}^\alpha(\partial\Omega_i, k)$ the closure in $B_{p,q}^\alpha(\Omega_i)$ of all smooth functions that vanish with order $l-2$ at the internal boundaries $\partial\Omega_i \cap \Omega$ and with order $k-1$ at the external boundaries $\partial\Omega_i \cap \partial\Omega$.

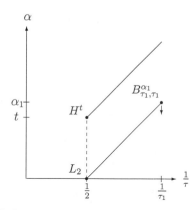

Figure 3.1: A classical DeVore-Triebel diagram for the embedding of Besov spaces with the L_2-embedding line $\frac{1}{\tau} = \frac{\alpha}{d} + \frac{1}{2}$ and H^t-embedding line $\frac{1}{\tau} = \frac{\alpha-t}{d} + \frac{1}{2}$.

Remark 3.1.3. *There are various alternative ways to introduce Besov spaces. A classical definition of Besov spaces on $\mathbb{R}^d$ used for instance in [116] is by a Fourier transform. Besov spaces on domains are then defined via restriction. A fairly general result showing the equivalence of this description with the one above, at least for bounded Lipschitz domains and the range of parameters of interest in this thesis, was shown in [56].*

3.2 Approximation with wavelet bases

In this section, we characterize the approximation spaces that correspond to the approximation of functions with wavelet bases. As a reference point and comparison, we will start with a classical *linear approximation* scheme.

3.2.1 Linear approximation with wavelet bases

In classical Galerkin methods, the elliptic equation $a(u, w) + \langle \mathcal{G}(u), w \rangle = f(w)$, $w \in \mathcal{H}$, is projected onto finite-dimensional linear subspaces X_j of $\mathcal{H}$. Thus, an approximation $u_j \in X_j$ to the solution $u \in \mathcal{H}$ is calculated by solving the reduced problems

$$a(u_j, v) + \langle \mathcal{G}(u_j), v \rangle = f(v), \quad w \in X_j \tag{3.4}$$

and letting j increase and X_j grow to obtain better estimates u_j for u. If the non-linearity $\mathcal{G}$ is a Nemitsky operator induced by a function $G \in C^1(\mathbb{R})$ with $G'(x) \geq 0$, $x \in \mathbb{R}$, it is shown in [127, Prop. 3.3] that the u_j are indeed near-best approximations from X_j in the sense that

$$\|u - u_j\|_{\mathcal{H}} \lesssim \inf_{v \in X_j} \|u - v\|_{\mathcal{H}}. \tag{3.5}$$

For linear problems, $\mathcal{G} = 0$, this result is known as Cea's lemma.

If we have available a wavelet basis $\Psi^{H_0^t(\Omega)} = \{\psi_\lambda\}_{\lambda \in \Lambda}$ for $\mathcal{H} = H_0^t(\Omega)$ constructed via a multiresolution analysis $\{V_j\}_{j \geq j_0}$, a natural choice for the spaces X_j is to set $X_j := V_j$. Bases for these spaces are simply given by $\Psi_{\leq j} := \{\psi_\lambda, |\lambda| \leq j\}$. This is called a *uniform* approach, because we simultaneously add all wavelets from the same level to the dictionary from which we calculate an approximation. The equivalent in finite element methods would be grid with a uniform mesh size over the entire domain. In view of the estimate (3.5), such classical uniform Galerkin methods realize an almost optimal approximation from the *linear* subspaces X_j, and are therefore interpreted as linear approximation schemes. With the help of the bases for V_j, Equation (3.4) can be reformulated as a *finite-dimensional* system in a similar way as in Section 2.3. From this system, the solutions u_j can be retrieved using, for instance, Newton's method. Therefore, to understand the convergence rate of classical Galerkin schemes, we have to understand linear approximation in $H_0^t(\Omega)$ with respect to the multiresolution spaces. The corresponding approximation spaces are again characterized as Sobolev spaces. This is shown in the following well-known result, see, e.g., [23, Ch. 3], which we present in the form from [123, Prop. 3.2].

Proposition 3.2.1. *Assume that a generalized Jackson inequality*

$$\inf_{v_j \in V_j} \|v - v_j\|_{H^s(\Omega)} \lesssim 2^{-j(t-s)} \|v\|_{H^t(\Omega)}, \quad v \in H^t(\Omega)$$

and a Bernstein inequality

$$\|v_j\|_{H^t(\Omega)} \lesssim 2^{j(t-s)} \|v_j\|_{H^s(\Omega)}, \quad v_j \in V_j$$

are valid for $0 \leq s \leq t < \gamma$. Then, we have the norm equivalence

$$\|v\|_{A_2^{sd}(H^t(\Omega))} \approx \|v\|_{H^{sd+t}(\Omega)}, \quad 0 \leq s \leq \frac{\gamma - t}{d}.$$

The Jackson and Bernstein estimates required here are somewhat more general than those needed in the previous section, but can be shown given sufficient Sobolev smoothness of the generator functions, see the proof of [76, Prop. 3.7]. For a function $v \in H^{sd+t}(\Omega)$ and $0 \leq s \leq \frac{\gamma - t}{d}$, Proposition 3.2.1 shows that it is $v \in A_2^{sd}(H^t(\Omega)) \subset A_\infty^{sd}(H^t(\Omega))$ and thus

$$E_n(v) \lesssim 2^{-sdj} \approx (\dim V_j)^{-s},$$

see also [123, Corollary 3.1]. In other words, to achieve a given rate $s > 0$ with respect to the degrees of freedom, classical linear approximation with wavelet bases requires that the function v we approximate has Sobolev regularity $v \in H^{sd+t}(\Omega)$. As we will explain in Section 3.5, in many cases, solutions to partial differential equations only have a very limited regularity in this scale. Therefore, we are now interested in whether other, nonlinear approximation schemes lead to possibly weaker regularity conditions to achieve a given rate with respect to the degrees of freedom.

Remark 3.2.2. *The connection between the convergence rate of uniform methods and the Sobolev regularity is not unique to approximation with wavelets. For uniform finite element methods, convergence estimates of the type*

$$\|u - u_h\|_{H^t(\Omega)} \lesssim h^{\alpha - t} \|u\|_{H^\alpha(\Omega)}$$

can be shown, where $u \in H_0^t(\Omega) \cap H^\alpha(\Omega)$ is the exact solution and $u_h \in H_0^t(\Omega)$ the approximation with finite elements of size h, see, e.g., [68] for details. The range of parameters $\alpha > t$ for which this estimate holds depends on the order of the elements. Since the overall number of finite elements behaves like h^{-d}, a convergence rate $s > 0$ with respect to the degrees of freedom can be guaranteed if $h^{\alpha - t} \|u\|_{H^\alpha(\Omega)} \lesssim (h^{-d})^{-s}$. This gives the condition $\alpha \geq sd + t$, which is the same as we have seen above in the case of uniform wavelet methods.

3.2.2 Nonlinear approximation with wavelet bases

A natural question to ask is how the approximation spaces can be characterized if we allow for a more flexible choice of the spaces X_j. Recall that in the previous subsection, we have studied approximation from the linear spaces V_j with a dimension of approximately 2^{dj}. Let us now instead consider the nonlinear spaces $X_j = \Sigma_{2^{dj}}$, where

$$\Sigma_n := \left\{ \sum_{\lambda \in \Lambda_n} c_\lambda \psi_\lambda, \ \Lambda_n \subset \Lambda, \ \#\Lambda_n \leq n \right\}.$$

is the set of all linear combinations of wavelets with at most n nonzero coefficients and $\Psi^{L_2(\Omega)} = \{\psi_\lambda\}_{\lambda \in \Lambda}$ is a wavelet basis for $L_2(\Omega)$. We denote the dual basis by $\tilde{\Psi}^{L_2(\Omega)}$

and assume that the primal and dual bases are constructed from a pair of biorthogonal MRA. The spaces $\Sigma_{2^{dj}}$ have approximately the same degrees of freedom as the spaces V_j. However, contrary to the latter spaces, we no longer prescribe the actual set of active wavelet indices, but only its maximum cardinality. Understood in this sense, approximation from the spaces $\Sigma_{2^{dj}}$ is indeed more flexible. The question to answer now is whether or not this flexibility pays off in terms of higher approximation rates. To this end, we have to characterize the corresponding approximation spaces. For this characterization, norm equivalences between Besov norms and weighted sequence norms of the expansion coefficients play a central role. Results of this kind for $\mathbb{R}^d$ were shown in [90, Ch. 6]. We present a version on bounded domains from [23, Th. 3.7.7].

Theorem 3.2.3. *Assume that the primal generator function ϕ is contained in $L_r(\Omega) \cap B^{\beta}_{p,q_0}(\Omega)$, with $0 < p \leq r$, arbitrary $q_0 \geq 0$ and $\beta > 0$, and the dual generator function $\tilde{\phi}$ is contained in $L_{r'}(\Omega)$ with $\frac{1}{r} + \frac{1}{r'} = 1$. In addition, assume that the primal MRA has l degrees of polynomial exactness, i.e., the spaces V_j reproduce polynomials up to order $l - 1$. Then, for $\alpha > 0$ with $d(\frac{1}{p} - \frac{1}{r}) < \alpha < \min\{\beta, l\}$, we have the norm equivalence*

$$\|v\|_{B^{\alpha}_{p,q}(\Omega)} \simeq \left(\sum_{j \geq j_0 - 1} 2^{qj(\alpha + d(\frac{1}{2} - \frac{1}{p}))} \Big(\sum_{|\lambda| = j} |c_\lambda|^p \Big)^{\frac{q}{p}} \right)^{\frac{1}{q}}, \tag{3.6}$$

where $v = \sum_{\lambda \in \Lambda} c_\lambda \psi_\lambda$.

Remark 3.2.4. *Note that in the special case $p = q = 2$, it is $B^{\alpha}_{2,2}(\Omega) = H^{\alpha}(\Omega)$ and the norm equivalence (3.6) corresponds to the norm equivalence (2.15) with Sobolev norms.*

This norm equivalence with Besov spaces is a key ingredient for the following characterization of the approximation spaces, which is a direct consequence of [23, Th. 4.2.2].

Theorem 3.2.5. *Assume that the spaces $B^{sd+t}_{\tau,\tau}(\Omega)$, $\frac{1}{\tau} = s + \frac{1}{2}$, admit a wavelet characterization (3.6) for $sd + t \in [t, t']$, $t' > t$. Then, the approximation spaces with respect to $X_j = \Sigma_{2^{dj}}$ are characterized as*

$$\|v\|_{A^{sd}_{\tau}(H^t(\Omega))} \simeq \|v\|_{B^{sd+t}_{\tau,\tau}(\Omega)}, \quad 0 < s < \frac{t' - t}{d}.$$

Recall that the nonlinear spaces X_j have 2^{dj} degrees of freedom. Hence, with the same arguments as in the previous subsection, we see that in order to achieve a given approximation rate s for the approximation of a function v in $H^t(\Omega)$, we now obtain the sufficient regularity condition $v \in B^{sd+t}_{\tau,\tau}(\Omega)$, $\frac{1}{\tau} = s + \frac{1}{2}$. Because τ is strictly smaller than 2, for arbitrary $\varepsilon > 0$, it is $B^{sd+t}_{\tau,\tau}(\Omega) \supset B^{sd+t-\varepsilon}_{2,2}(\Omega) = H^{sd+t-\varepsilon}(\Omega)$, where the latter was the space that guaranteed an approximation rate arbitrarily close to

s when using linear approximation. This shows that, in this context, the conditions to achieve a given rate s are indeed weaker when using nonlinear approximation schemes instead of linear approximation. We will see in Section 3.5 that for the problems we are interested in, the regularity measured in the above Besov scale is indeed substantially higher than in the classical Sobolev scale. In other words, we can hope for a significantly better rate when considering nonlinear approximation.

On the other hand, since we are interested in finding the solution of boundary value problems, the object we approximate is unknown. Therefore, it is usually not possible to calculate its best N-term approximation. We have seen in the beginning of Subsection 3.2.1, that for linear approximation schemes in the spaces V_j, at least a *near-best approximation* can be realized in the sense that we only lose a constant compared to the best approximation. Constructing an algorithm that realizes the rate of the nonlinear best N-term approximation, however, is significantly more challenging. This is because, for any N, it requires to choose a near-optimal set of N active indices from the infinite index set Λ. To do so, it is necessary to have some information about the solution we want to approximate. Since we do not want to rely on a-priori information, we have to use the information that is gradually gained during the process of approximating the solution in order to identify the most relevant coefficients and to adapt the set of active coefficients appropriately. Hence, to realize the rate of the best N-term approximation, it is necessary to work with *adaptive algorithms*. The design of such algorithms will be the focus of Chapter 4. There, the rate of nonlinear approximation will serve as a benchmark for the algorithms we want to design. Before coming to this, however, it will be helpful to relocate the problem of nonlinear approximation to a problem in $\ell_2(\Lambda)$, the space in which the discretized equations are formulated.

3.3 Unrestricted nonlinear approximation in ℓ_2

In Section 2.3, we have seen that the discretization of a semilinear PDE with the help of a wavelet basis or frame leads to an equivalent problem in $\ell_2(\Lambda)$. Therefore, it will be helpful to discuss nonlinear approximation in this sequence space. This relocation of the approximation problem is feasible, because the Riesz basis property (2.1) implies that with a Riesz basis $\Psi = \Psi^{H_0^t(\Omega)}$ for $H_0^t(\Omega)$, it holds that

$$\|v - w\|_{H^t(\Omega)} \eqsim \|\mathbf{v} - \mathbf{w}\|_{\ell_2(\Omega)}, \quad \mathbf{v}, \mathbf{w} \in \ell_2(\Lambda), \, v = \mathbf{v}^\top \Psi, w = \mathbf{w}^\top \Psi.$$

Hence, the distance between two functions in $H_0^t(\Omega)$ is, up to constants, the same as the distance of their expansion coefficients in the Riesz basis. In the case of a frame, some care has to be taken due to the redundancy of the system. However, as we see in Estimate (2.5), an analogous equivalence holds if we consider the minimal difference between all possible expansion coefficients of v and w.

At first, we focus on unrestricted nonlinear approximation. In the subsequent section, we will consider approximation with respect to tree structures. Let us start by introducing the basic terms and definitions, roughly following [25]. For further reading and proofs, we also refer to the surveys [23, 52].

3.3.1 N-term approximation

Let Λ be a countable index set. Typically, we think of Λ as the index set belonging to a wavelet basis or frame. We define

$$\Sigma_{N,\ell_2(\Lambda)} := \{\mathbf{w} \in \ell_2(\Lambda), \#\operatorname{supp}\mathbf{w} \leq N\}$$

as the space of all vectors indexed by Λ with at most N nonzero entries. For $\mathbf{v} \in \ell_2(\Lambda)$, the *error of the best N-term approximation* is given by

$$\sigma_N(\mathbf{v}) := \inf_{\mathbf{w}\in\Sigma_{N,\ell_2(\Lambda)}} \|\mathbf{v} - \mathbf{w}\|_{\ell_2(\Lambda)}.$$

A vector $\mathbf{w} \in \ell_2(\Lambda)$, in which the infimum is taken as a minimum, is called *best N-term approximation* to $\mathbf{v}$. Such a vector can be constructed by simply taking the N entries of $\mathbf{v}$ with largest absolute value. Since several entries can have the same absolute value, this best N-term approximation is not unique in general. This, however, is not a serious problem because in the following, we will only work with the error of the best N-term approximation. To collect all vectors for which this error decays with a common rate $s > 0$, we set

$$\|\mathbf{v}\|_{\mathcal{A}^s} := \sup_{N\geq 0}(N+1)^s\sigma_N(\mathbf{v}), \quad \mathbf{v} \in \ell_2(\Lambda)$$

with $\sigma_0(\mathbf{v}) := \|\mathbf{v}\|_{\ell_2(\Lambda)}$, and we define the discrete approximation class

$$\mathcal{A}^s := \{\mathbf{v} \in \ell_2(\Lambda), \|\mathbf{v}\|_{\mathcal{A}^s} < \infty\}.$$

By definition, we have $\mathbf{v} \in \mathcal{A}^s$ if and only if $\sigma_N(\mathbf{v}) \lesssim N^{-s}$. An alternative characterization of these classes can be established with the help of the *weak ℓ_τ-spaces* we introduce now. First, for a vector $\mathbf{v}$, we denote by v_k^* its k-th largest coefficient in absolute value. Then, we define $\ell_\tau^w(\Lambda)$, $0 < \tau < 2$ as the space of all vectors $\mathbf{v} \in \ell_2(\Lambda)$, for which

$$|\mathbf{v}|_{\ell_\tau^w(\Lambda)} := \sup_{k\geq 1} k^{1/\tau}|v_k^*|$$

is finite. On this space, we can define the quasi-norm

$$\|\mathbf{v}\|_{\ell_\tau^w(\Lambda)} := \|\mathbf{v}\|_{\ell_2(\Lambda)} + |\mathbf{v}|_{\ell_\tau^w(\Lambda)}.$$

The name *weak ℓ_τ-space* is justified by the embeddings

$$\ell_\tau(\Lambda) \subset \ell_\tau^w(\Lambda) \subset \ell_{\tau+\varepsilon}(\Lambda), \quad 0 < \tau < \tau + \varepsilon < 2. \tag{3.7}$$

The following characterization of the approximation classes was stated in [25, Prop. 3.2], see also [52, Ch. 5] for a short proof.

Proposition 3.3.1. *Let $s > 0$ and $\frac{1}{\tau} = s + \frac{1}{2}$. Then, we have*

$$\|\mathbf{v}\|_{\ell_\tau^w(\Lambda)} \approx \|\mathbf{v}\|_{\mathcal{A}^s}$$

with constants only depending on τ when τ goes to zero.

3.3.2 N-term approximation and wavelet Riesz bases

With the help of the above concepts, we are now ready to link the smoothness of a function to the convergence rate of the best N-term approximation with respect to a given wavelet basis. Let us start with the case of $\Psi^{L_2(\Omega)}$ being a Riesz basis for $L_2(\Omega)$ consisting of compactly supported wavelets, diam supp $\psi_\lambda \eqsim 2^{-|\lambda|}$, such that a rescaled version $\Psi^{H^t(\Omega)} = \mathbf{D}^{-t}\Psi^{L_2(\Omega)}$ is a Riesz basis for $H^t(\Omega)$. Then, we can establish the following equivalence, see [25, 123].

Proposition 3.3.2. *Assume that $\Psi^{L_2(\Omega)}$ fulfills a norm equivalence (3.6) and let $v = \mathbf{c}^\top \Psi^{H^t(\Omega)} = (\mathbf{D}^{-t}\mathbf{c})^\top \Psi^{L_2(\Omega)}$. Then, the following two statements are equivalent.*

(i) $v \in B^{sd+t}_{\tau,\tau}(\Omega)$, $\frac{1}{\tau} = s + \frac{1}{2}$,

(ii) $\mathbf{c} \in \ell_\tau(\Lambda)$.

Proof. The proof is an application of the norm equivalences. Note that the expansion coefficients of v with respect to $\Psi^{L_2(\Omega)}$ are given by $(2^{-t|\lambda|}c_\lambda)_{\lambda \in \Lambda}$. Hence, by (3.6), it is

$$\|v\|_{B^{sd+t}_{\tau,\tau}(\Omega)} \eqsim \left(\sum_{j \geq j_0-1} 2^{\tau j(sd+t+d(\frac{1}{2}-\frac{1}{\tau}))} \sum_{|\lambda|=j} 2^{-jt\tau}|c_\lambda|^\tau \right)^{\frac{1}{\tau}}$$

$$= \left(\sum_{j \geq j_0-1} \sum_{|\lambda|=j} |c_\lambda|^\tau \right)^{\frac{1}{\tau}}$$

$$= \|\mathbf{c}\|_{\ell_\tau(\Lambda)}.$$

$\square$

Let us shortly comment on this result. It states that if a function has sufficient Besov smoothness $v \in B^{sd+t}_{\tau,\tau}(\Omega)$, its expansion coefficients with respect to a suitable wavelet basis for $H^t(\Omega)$ are contained in $\ell_\tau(\Lambda)$. The latter space, in turn, is contained in $\ell_\tau^w(\Lambda) = \mathcal{A}^s$. This means that we expect the best N-term approximation to converge with rate s in $H^t(\Omega)$. The range for which this implication holds is governed by the range of parameters for which the norm equivalence (3.6) is valid. Recall that this range is limited by $sd + t \leq \gamma := \min\{\alpha, l\}$ with the notation from Theorem 3.2.3. Given that the Besov regularity of the generators is sufficiently large, this means that the highest rate we can expect is limited by $s \leq s^* := \frac{l-t}{d}$, even if we approximate a function with arbitrarily high Besov regularity. Note that the upper bound s^* only depends on the order l of the wavelets, the order t of the Sobolev space in which we measure the approximation error and the space dimension. In particular, if $v \in B^{sd+t}_{\tau,\tau}(\Omega)$ can be shown for a large rate of parameters s, we can increase the convergence rate of the best N-term approximation by using wavelet bases with a higher number of vanishing moments.

Remark 3.3.3. *As outlined before, in our practical applications we are mostly concerned with spaces with incorporated boundary conditions. In [23, Section 3.10], see also [76, 123], it is outlined that an analogous norm equivalence as in Theorem 3.2.3, and thus a characterization as in Proposition 3.3.2, also holds for the wavelet bases that characterize $H_0^t(\Omega)$ instead of $H^t(\Omega)$. To explain this, let $v = \mathbf{c}^\top \Psi^{H_0^t(\Omega)} = (\mathbf{D}^{-t}\mathbf{c})^\top \Psi^{L_2(\Omega)}$. Under the additional restriction $sd + t - \frac{1}{\tau} \notin \{0,\dots,l-2\}$, we then have $v \in B_{\tau,\tau}^{sd+t}(\partial\Omega, t)$ if and only if $\mathbf{c} \in \ell_\tau(\Lambda)$, $\frac{1}{\tau} = s + \frac{1}{2}$. This equivalence holds in the same range $0 < s < \frac{\gamma-t}{d}$.*

3.3.3 N-term approximation and wavelet frames

We now describe what differences occur if the underlying generating system is only a wavelet frame as constructed in Subsection 2.2.5 and no longer a basis. Using such a possibly overcomplete generating system has the consequence that the expansion coefficients of a given function are no longer unique in general. Hence, if they are chosen arbitrarily among all possible choices, we cannot expect to be able to relate the smoothness of a function to the convergence rate of the best N-term approximation of these coefficients. However, if we choose a particular representation that is created via a smooth partition of unity, a result analogous to Proposition 3.3.2 can be shown. To formulate this result, let $\Psi^{L_2(\Omega)}$ be the wavelet frame from Proposition 2.2.16 with the particular dual frame $\tilde{\Psi}^{L_2(\Omega)}$ and $\Psi^{H_0^t(\Omega)} = \mathbf{D}^{-t}\Psi^{L_2(\Omega)}$ be the rescaled frame in $H_0^t(\Omega)$. Moreover, assume that the bases from which the aggregated frame is composed fulfill the norm equivalence from Remark 3.3.3 for all $0 < s < \frac{l-t}{d}$.

Proposition 3.3.4 ([123, Prop. 3.5]). *Assume that there exists a smooth partition of unity $\{\sigma_i\}_{i=0}^{m-1}$ with respect to the overlapping domain decomposition $\{\Omega_i\}_{i=0}^{m-1}$ of Ω. Let $v \in H_0^t(\Omega) \cap B_{\tau,\tau}^{sd+t}(\Omega)$, $0 < s < \frac{l-t}{d}$, $\frac{1}{\tau} = s + \frac{1}{2}$ and $sd + t - \frac{1}{\tau} \notin \{0,\dots,l-2\}$. Then, the expansion coefficients $\mathbf{c} \in \ell_2(\Lambda)$, $\mathbf{c}^\top \Psi^{H_0^t(\Omega)} = v$, calculated with scaled dual frame $\mathbf{D}^t\tilde{\Psi}^{L_2(\Omega)}$ from Proposition 2.2.16,*

$$c_\lambda = \langle v, 2^{t|\lambda|}\sigma_i\tilde{\psi}_{(i,\lambda)}\rangle, \quad \lambda \in \{i\} \times \Lambda_i, \tag{3.8}$$

are contained in $\ell_\tau(\Lambda)$.

Proof. Because the σ_i are smooth and a partition of unity, from [116, Section 2.8.2] and the property (iii) in Definition 2.2.13, it follows that $\sigma_i v \in B_{\tau,\tau}^{sd+t}(\partial\Omega_i, t)$. Hence, by Remark 3.3.3, it holds that

$$\mathbf{c}^{(i)} := (\langle v, 2^{|\lambda|}\sigma_i\tilde{\psi}_{(i,\lambda)}\rangle)_{\lambda\in\{i\}\times\Lambda_i} \in \ell_\tau(\{i\} \times \Lambda_i).$$

Therefore, we have $\mathbf{c} = (\mathbf{c}^{(0)},\dots,\mathbf{c}^{(m-1)}) \in \ell_\tau(\Lambda)$. $\qquad\square$

Unfortunately, for many cases of interest such as the L-shaped domain with the standard overlapping domain decomposition (2.18), the above result does not apply. That is because in such cases there is no smooth partition of unity, see Example 2.2.15.

Under slightly stronger assumptions, however, a similar characterization as in Proposition 2.2.16 was shown in [33]. We will outline these results now. Hence, for the remainder of this subsection, let Ω be the L-shaped domain.

Looking at the proof of Proposition 3.3.4, the key task is to establish Besov regularity for the weighted functions $\sigma_i v$. From the construction in Example 2.2.15, we know that the σ_i are smooth except at the origin. This singularity can, in a way, be compensated for if the function v decays fast enough around the origin and has a slightly higher Sobolev regularity than what is guaranteed by the embedding $B^{sd+t}_{\tau,\tau}(\Omega) \subset H^t(\Omega)$. This motivates the following assumption.

Assumption 3.3.5. *Assume that there exists a $\vartheta > 0$ such that v is contained in $H^{t+\vartheta}(\Omega) \cap H^t_0(\Omega)$ and that for all multiindices α with $|\alpha| = j$, $0 \leq j \leq t$, there exist $\beta_j > t - (j+1)$ such that*

$$D^\alpha v(x,y) = \mathcal{O}(r(x,y)^{\beta_j}), \quad r(x,y) \to 0,$$

where $(r(x,y), \theta(x,y))$ denotes the polar coordinates with respect to the origin.

Under this assumption, Besov regularity of the functions $\sigma_i v$ can be established.

Lemma 3.3.6 ([33, Lemma 4.7]). *Let Assumption 3.3.5 hold, let $\alpha > t$, $\delta \in (0, \alpha - t)$ such that $v \in B^\alpha_{\tau,\tau}(\Omega)$, $\frac{1}{\tau} = \frac{\alpha - (t+\delta)}{2} + \frac{1}{2}$. Then, for any $\varepsilon \in (0, 2+\delta)$, $\alpha' \in [1-\varepsilon, \alpha)$ it is $\sigma_i v \in B^{\alpha'}_{\tau',\tau'}(\Omega)$, $i = 0,1$, $\frac{1}{\tau'} = \frac{\alpha' - (1-\varepsilon)}{2} + \frac{1}{2}$.*

This is one of the main ingredients for the following result, which gives us the desired relation between a function v and the decay of its expansion coefficients.

Proposition 3.3.7 ([33, Th. 4.5]). *Let the conditions from Lemma 3.3.6 be satisfied and let*

$$\eta^* := \min\{\vartheta, 1 - t + \min_{0 \leq j \leq t}(j + \beta_j)\}. \tag{3.9}$$

Then, the expansion coefficients $\mathbf{c} \in \ell_2(\Lambda)$, $\mathbf{c}^\top \Psi = v$ from Equation (3.8), i.e., the expansion coefficients with respect to the scaled dual frame $\mathbf{D}^t \tilde{\Psi}^{L_2(\Omega)}$, are contained in $\ell_{\tau_0}(\Lambda)$, where $\frac{1}{\tau_0} = \frac{\alpha' - t}{2} + \frac{1}{2}$, for all

$$t < \alpha' < \min\{d, \frac{\eta^*\alpha + t - 1}{t + \eta^* - 1} + t - 1\}.$$

Let us shortly interpret this result. In classical N-term approximation with wavelet bases on $\Omega \subset \mathbb{R}^2$ or with frames constructed via a smooth partition of unity, the spaces characterizing the approximation rate s are the spaces $B^{2s+t}_{\tau,\tau}(\Omega)$, $\frac{1}{\tau} = s + \frac{1}{2}$. Written in the usual adaptivity scale, these are the spaces $B^\alpha_{\tau,\tau}(\Omega)$, $\frac{1}{\tau} = \frac{\alpha - t}{2} + \frac{1}{2}$. When dealing with wavelet frame approximation on the L-shaped domain, the corresponding embedding line is slightly altered to $\frac{1}{\tau} = \frac{\alpha - (t+\delta)}{2} + \frac{1}{2}$ with a fixed, but arbitrarily small $\delta > 0$. Hence, to end up with the same rate of the best N-term approximation, we need to start with a function that has only a slightly higher Besov regularity. This is best illustrated with the help of a DeVore-Triebel diagram as in Figure 3.2, see also [33]. In practice, the above result has the consequence that the optimal approximation rates are mostly the same when using an aggregated wavelet frame instead of a Riesz basis on the L-shaped domain, although the former is much easier to construct than the latter.

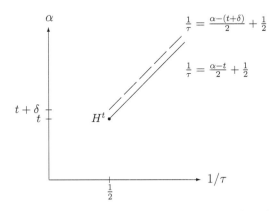

Figure 3.2: A DeVore-Triebel diagram showing the embedding lines governing the rate of the best N-term approximation for bases (solid) and for frames on the L-shaped domain (dashed).

3.4 Tree-structured approximation in ℓ_2

When dealing with nonlinear problems of the form (2.21), at some point we will have to evaluate nonlinear expressions in wavelet coordinates. As we will see later, such an evaluation can only be constructed efficiently if the set of active wavelet indices we work with is of the form of a so-called *tree*.

To give a rough intuition of why this is the case, recall that the handling of nonlinear terms requires the approximation of the expressions $\langle \mathcal{G}(\mathbf{v}^\top \Psi), \psi_\mu \rangle$, at least for indices μ on a subset $\mathcal{T}$ of Λ containing the most relevant entries. The naive approach would be to use integration on a sufficiently fine grid. This, however, is very costly. An alternative idea is to use the two-scale relation (2.10) to subsequently relocate the problem in a top-down procedure to the next lower level all the way to the lowest level. Such an approach requires that the set of indices we work on is structured, i.e., that the required indices on the next lower level are contained in $\mathcal{T}$. It will turn out in Subsection 4.2.3 that if we work on tree structures, this is indeed the case.

The approximate evaluation of linear and nonlinear functionals using such trees was, among others, developed and analyzed in [7, 10, 24, 27, 28, 47]. These concepts were generalized to aggregated wavelet frames in [75, 76]. Following these references, we introduce tree structures and collect some important results on N-term approximation with respect to this structural constraint. In particular, we will note that the spaces ensuring a given approximation rate are again Besov spaces and that only slightly stricter regularity assumptions are needed to achieve the same rates as with unrestricted N-term approximation.

3.4.1 Trees and aggregated trees

We start by explaining what is meant by tree-structured index sets. To understand the basic ideas, it is easiest to start with classical wavelet bases $\Psi^{L_2(\mathbb{R})} = \{\psi_{j,k}, j, k \in \mathbb{Z}\}$ for $L_2(\mathbb{R})$ constructed by dilation and translation of a single function ψ, i.e.,

$$\psi_{j,k}(x) = 2^{j/2}\psi(2^j x - k), \quad j, k \in \mathbb{Z}.$$

If, as in the case of the Haar wavelet (2.7), ψ is supported in $[0, 1]$, then $\psi_{j,k}$ is supported in $2^{-j}[k, k+1]$. On the next higher level, the supports of $\psi_{j+1,2k}$ and $\psi_{j+1,2k+1}$ are contained in the support of $\psi_{j,k}$. Hence, we call $(j+1, 2k)$ and $(j+1, 2k+1)$ *children* of (j, k). Conversely, (j, k) is referred to as the *parent* of $(j+1, 2k)$ and $(j+1, 2k+1)$. Such a relation between the indices can be generalized to wavelet indices with a higher difference of levels: An index (j, k) will be called *successor* of (j', k'), $j > j'$ if the support of $\psi_{j,k}$ is contained in the support of $\psi_{j',k'}$, in short $(j, k) \succ (j', k')$. This induces a partial order $\succ$ on the set of indices. In a similar manner, it is stated in [27] that such a structure can be generalized to all kinds of isotropic wavelet bases indices. We will outline below how a practical construction for bases on the unit cube can be done. The existence of such a partial order allows us to give the following definition:

Definition 3.4.1. *Assume that we have a partial order $\succ$ on a set $\mathcal{J}$ of indices. A finite subset $\mathcal{T} \subset \mathcal{J}$ is called a* tree, *if for each $\mu \in \mathcal{T}$, all its* predecessors, *i.e., all $\lambda \in \mathcal{J}$ with $\mu \succ \lambda$, are contained in $\mathcal{T}$.*

Remark 3.4.2. *The restriction to finite index sets might seem confusing in view of the basis on the real line, because in this case, each index has infinitely many predecessors. However, for the constructions on bounded domains that we are interested in, there exists a minimum level of the wavelet indices and each index has only finitely many predecessors.*

Since in our applications, the index set $\mathcal{J}$ belongs to a wavelet basis, or, as we will see later, the generators from which the basis is constructed, to each element $\lambda \in \mathcal{J}$, a level $|\lambda|$ can be assigned. This allows us to give the following definitions, that will be needed later. In particular, this generalizes the above notions of parent and child indices to such index sets.

Definition 3.4.3. *Let $\mathcal{T} \subset \mathcal{J}$ be a tree. If $\lambda \in \mathcal{T}$ is a successor of $\mu \in \mathcal{T}$ with $|\lambda| - |\mu| = 1$, then μ is a* parent *of λ, and λ is called a* child *of μ. Two indices with the same parent are called* siblings. *The set of all children of μ is denoted by $\mathcal{C}(\mu)$. An index $\lambda \in \mathcal{J}$ is called* outer leave, *if it does not belong to $\mathcal{T}$, but its parents do. The set of outer leaves of $\mathcal{T}$ is denoted by $\mathcal{D}(\mathcal{T})$.*

Let us now sketch how tree structures can be defined for wavelet bases on the unit cube $\square = (0,1)^d$ indexed by $\Lambda^\square$ as constructed in Subsection 2.2.4. There, the situation is a little more complicated than on the real line, because not all wavelets are generated only by dilation and translation of a single function. Hence, to describe the construction, following [6, 76], we will introduce another structural tool, namely *reference cubes*. To define such cubes, we first have to fix some notation. For a wavelet basis index $\lambda = (j, \mathbf{e}, \mathbf{k}) \in \Lambda^\square$, we denote by $\lambda^\circ := (j, \mathbf{k})$ the corresponding *generator index*. We write $|\lambda^\circ| := j$, and for a collection $U \subset \Lambda^\square$, we set $U^\circ := \{\lambda^\circ, \lambda \in U\}$. To each generator index λ°, we then assign a closed reference cube $\square_{\lambda^\circ}$ with the following properties:

- For each $\lambda \in \Lambda^\square$, it is $\square_{\lambda^\circ} \subset \operatorname{supp} \psi_\lambda$.

- For $|\lambda| = |\mu|$ with $\lambda^\circ \neq \mu^\circ$, $\square_{\lambda^\circ} \cap \square_{\mu^\circ}$ is either empty or a lower-dimensional face of both $\square_{\lambda^\circ}$ and $\square_{\mu^\circ}$.

- For each $j \geq j_0$, it is $\overline{\square} = \bigcup_{\lambda \in \Delta_j} \square_{\lambda^\circ}$.

- For each λ°, there are indices μ_s°, $s = 1, \ldots, N$, $N \geq 2$ on the next higher level such that $\square_{\lambda^\circ} = \bigcup_{s=1}^N \square_{\mu_s^\circ}$, and N is uniformly bounded. Typically, it is $N = 2^d$.

With the help of the reference cubes, we are ready to define a partial order on the set $I^\square := \{(j, \mathbf{k}), j \geq j_0, \mathbf{k} \in \Delta_j\}$ of generator indices. For two such indices $\lambda^\circ, \mu^\circ \in I^\square$, we say that λ° a successor of μ°, $\lambda^\circ \succ \mu^\circ$, if $\square_{\lambda^\circ} \subsetneq \square_{\mu^\circ}$. This induces a tree structure on $\mathcal{J} := I^\square$ in the sense of Definition 3.4.1. Moreover, the reference cubes allow us to visualize tree-structured index sets, at least on the unit interval. In this case, the reference cubes are simply subintervals. In Figure 3.3, c.f. [76], the reference cubes are drawn as rectangles. We see an unstructured index set on the left and the smallest tree containing it in the center. On the right, we see the smallest *complete* tree containing the indices, see the Definition 3.4.4 below.

So far, we have only defined trees on the set $I^\square$ of generator indices. However, the concept can be generalized to subsets $\mathcal{T}$ of $\Lambda^\square$, the set corresponding to the wavelet basis. This is simply done by saying that $\mathcal{T} \subset \Lambda^\square$ has a *tree-like structure*, if the corresponding set $\mathcal{T}^\circ$ is a tree in the above sense. For simplicity, such a set $\mathcal{T}$ will also be called a tree.

We have indicated above that we will use a top-down procedure to deal with the evaluation of nonlinear problems. Hence, the indices in $I^\square$ on the lowest level will play a special role. Therefore, an index $\lambda^\circ \in I^\square$ is called *root*, if $|\lambda^\circ| = j_0$. Analogously,

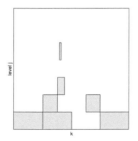 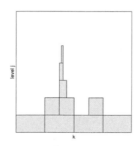 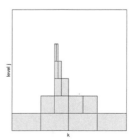

Figure 3.3: Unstructured index set (left), the smallest tree (center) and the smallest complete tree (right) containing it.

the indices in the set $\Lambda^\square$ on the lowest level are also called roots and the set of all roots is denoted by

$$\mathcal{R} := \{\lambda \in \Lambda^\square,\ |\lambda| = j_0\}.$$

In the algorithms, it will be useful if the trees we work with contain all roots and fulfill additional properties that are defined as follows.

Definition 3.4.4. • *A tree $\mathcal{T}^\circ \subset I^\square$ is called* complete *if it contains all roots and if for each $\lambda^\circ \in \mathcal{T}^\circ$, all its siblings also belong to $\mathcal{T}^\circ$.*

• *A tree $\mathcal{T} \subset \Lambda^\square$ is called* complete *if $\mathcal{T}^\circ$ is complete and if $\lambda = (j, \mathbf{e}, \mathbf{k}) \in \mathcal{T}$ implies that all $\mu = (j', \mathbf{e}', \mathbf{k}') \in \Lambda^\square$ with $j = j'$ and $\mathbf{k} = \mathbf{k}'$ also belong to $\mathcal{T}$.*

Assuming complete trees, however, does not pose a serious restriction. This is because by the uniform boundedness of the number of siblings, for a non-empty tree $\mathcal{T}$ there exists a complete tree $\mathcal{T}_c$ containing $\mathcal{T}$ such that

$$\#\mathcal{T}_c \lesssim \#\mathcal{T},$$

see [76, Section 3.3.1]. Hence, we will assume all trees to be complete in the following.

The definition of tree structures can easily be generalized to aggregated wavelet frames constructed along the lines of Subsection 2.2.5. For simplicity, we assume that the Riesz bases on the subdomains share the index set $\Lambda_i = \Lambda^\square$ from the basis on the cube.

Definition 3.4.5. *A set*

$$\mathcal{T}^\circ = \bigcup_{i=0}^{m-1} \{i\} \times \mathcal{T}_i^\circ \subset \bigcup_{i=0}^{m-1} \{i\} \times I^\square$$

is called (aggregated) tree *if all the $\mathcal{T}_i^\circ$ are trees in the sense of Definition 3.4.1. In the same fashion*

$$\mathcal{T} = \bigcup_{i=0}^{m-1} \{i\} \times \mathcal{T}_i \subset \Lambda = \bigcup_{i=0}^{m-1} \{i\} \times \Lambda_i$$

is called (aggregated) tree *if all the $\mathcal{T}_i$ are trees. In the above cases, $\mathcal{T}_i^\circ$ and $\mathcal{T}_i$ are called* local trees. *An aggregated tree is* complete *if all its local trees are complete.*

3.4.2 *N*-term tree approximation and wavelet Riesz bases

We are now ready to define the tree-wise analog to the approximation classes $\mathcal{A}^s$ from Subsection 3.3.1. We start with the case of wavelet bases, roughly following [27, 28]. Since the bases we work with typically are transformed versions of bases on the unit cube, we write $\Lambda^\square$ for the underlying index set. First, let

$$\Sigma_{N,\mathcal{T}} := \{\mathbf{v} \in \ell_2(\Lambda^\square),\ \#\operatorname{supp}\mathbf{v} \leq N,\ \operatorname{supp}\mathbf{v} \text{ is a tree}\}.$$

Now define the *error of the best N-term tree approximation* to $\mathbf{v} \in \ell_2(\Lambda^\square)$ as

$$\sigma_{N,\mathcal{T}}(\mathbf{v}) := \inf_{\mathbf{w} \in \Sigma_{N,\mathcal{T}}} \|\mathbf{v} - \mathbf{w}\|_{\ell_2(\Lambda^\square)}$$

when N is larger or equal than the number $\#\mathcal{R}$ of roots and $\sigma_{N,\mathcal{T}}(\mathbf{v}) := \|\mathbf{v}\|_{\ell_2(\Lambda^\square)}$ otherwise. This minor modification is due to assuming completeness of the trees we work with. As before, a vector $\mathbf{w} \in \Sigma_{N,\mathcal{T}}$ in which the infimum is taken is called *best N-term tree approximation* to $\mathbf{v}$. Finally, the spaces $\mathcal{A}_{\mathcal{T}}^s$, $s > 0$ are defined as the sets of all vectors $\mathbf{v} \in \ell_2(\Lambda^\square)$ for which

$$\|\mathbf{v}\|_{\mathcal{A}_{\mathcal{T}}^s} := \sup_{N \geq 0}(N+1)^s \sigma_{N,\mathcal{T}}(\mathbf{v})$$

is finite. This quantity is a quasi-norm on $\mathcal{A}_{\mathcal{T}}^s$. By definition, the spaces $\mathcal{A}_{\mathcal{T}}^s$ consist of all the vectors whose best N-term *tree* approximation converges with rate s. Clearly, $\mathcal{A}_{\mathcal{T}}^s$ is contained in $\mathcal{A}^s$. An important question to be answered is how much smaller this space actually is, in other words, how the approximation rate suffers from the additional restriction of approximating with tree-structured vectors only. To answer this question, it will be helpful to resort to a modified type of *weak ℓ_τ-spaces*. In the context of tree approximation, their definition is somewhat different from the case of unrestricted index sets as in Subsection 3.3.1. To begin with, for a vector $\mathbf{v} = (v_\lambda)_{\lambda \in \Lambda^\square} \in \ell_2(\Lambda^\square)$, we define

$$\tilde{v}_\lambda := \left(\sum_{\mu \succeq \lambda} |v_\mu|^2\right)^{1/2}$$

as the part of the ℓ_2-norm of $\mathbf{v}$ belonging to λ and all its successors, where $\mu \succeq \lambda$ means that $\mu \succ \lambda$ or $\mu = \lambda$. We then set $\tilde{\mathbf{v}} := (\tilde{v}_\lambda)_{\lambda \in \Lambda^\square}$. The *weak ℓ_τ-tree-space* $\ell_{\tau,\mathcal{T}}^w(\Lambda^\square)$ is now defined as the space of all vectors $\mathbf{v} \in \ell_2(\Lambda^\square)$, for which

$$\|\mathbf{v}\|_{\ell_{\tau,\mathcal{T}}^w(\Lambda^\square)} := \|\tilde{\mathbf{v}}\|_{\ell_\tau^w(\Lambda^\square)}$$

is finite. The space $\ell_{\tau,\mathcal{T}}^w(\Lambda^\square)$ is contained in $\mathcal{A}_{\mathcal{T}}^s$, $\frac{1}{\tau} = s + \frac{1}{2}$, but the converse inclusion does not hold in general, see [28, Section 2.3]. Recall that in the case of unrestricted approximation in ℓ_2, the corresponding spaces are actually equal. With the help of these spaces, however, we can now establish the connection between the Besov regularity of a function and the convergence rate of the best N-term tree approximation of its expansion coefficients.

Proposition 3.4.6 ([28, Remark 2.3])**.** *Let the conditions from Proposition 3.3.2 be fulfilled and $v = \mathbf{c}^\top \Psi^{H^t(\Omega)} \in B_{\tau',q}^{sd+t}(\Omega)$ for some $\frac{1}{\tau'} < \frac{1}{\tau} = s + \frac{1}{2}$ and some $0 < q \leq \infty$. Then, it is $\mathbf{c} \in \ell_{\tau,\mathcal{T}}^w(\Lambda^\square)$.*

Let us shortly bring this result into line with the preceding results. To guarantee that the expansion coefficients of a function v are contained in $\ell_{\tau,\mathcal{T}}^w(\Lambda^\square) \subset \mathcal{A}_{\mathcal{T}}^s$, we need $v \in B_{\tau',q}^{sd+t}(\Omega)$ with $\tau' > \tau = \left(s + \frac{1}{2}\right)^{-1}$. Comparing this with the result in Proposition 3.3.2, we see that regularity is now measured in a slightly stronger Besov metric. The difference between τ' and τ, however, is arbitrarily small. Hence, we may expect almost the same rate when restricting ourselves to tree approximation as compared to unrestricted approximation.

3.4.3 N-term aggregated tree approximation and wavelet frames

So far, we have outlined the properties of N-term approximation with respect to trees. Following [75, 76], we are now going to describe how these concepts can be extended to aggregated trees corresponding to aggregated wavelet frames.

Similarly to the previous subsection, we define $\Sigma_{N,\mathcal{AT}}$ as the nonlinear space of all vectors in $\ell_2(\Lambda)$ in aggregated tree structure with at most N nonzero entries. Then, we set

$$\sigma_{N,\mathcal{AT}}(\mathbf{v}) := \inf_{\mathbf{w} \in \Sigma_{N,\mathcal{AT}}} \|\mathbf{v} - \mathbf{w}\|_{\ell_2(\Lambda)}, \quad \mathbf{v} \in \ell_2(\Lambda),$$

as before with the modification $\sigma_{N,\mathcal{AT}}(\mathbf{v}) := \|\mathbf{v}\|_{\ell_2(\Lambda)}$ if N is smaller than the number of roots. Finally, we can define the approximation space

$$\mathcal{A}_{\mathcal{AT}}^s := \{\mathbf{v} \in \ell_2(\Lambda), \|\mathbf{v}\|_{\mathcal{A}_{\mathcal{AT}}^s} := \sup_{N \geq 0}(N+1)^s \sigma_{N,\mathcal{AT}}(\mathbf{v}) < \infty\}$$

of all vectors in $\ell_2(\Lambda)$ whose best N-term approximation in aggregated tree structure converges with a common rate s. In a straightforward manner, these spaces are related to the tree approximation spaces from the previous subsection. To explain this, note first that every $\mathbf{v} \in \ell_2(\Lambda)$ can be rearranged as $\mathbf{v} = (\mathbf{v}_0, \ldots, \mathbf{v}_{m-1}) \in \ell_2(\Lambda)$ with $\mathbf{v}_i \in \{i\} \times \Lambda^\square$. Then, it is easy to see that $\mathbf{v}$ is contained in $\mathcal{A}_{\mathcal{AT}}^s$ if and only if all the $\mathbf{v}_i$ are contained in $\mathcal{A}_{\mathcal{T}}^s$, see also [76, Lemma 3.4.1]. Moreover, the approximation spaces based on aggregated trees are again related to certain modified weak ℓ_τ-spaces. Directly adapting the definition of weak tree ℓ_τ-spaces, we set

$$\ell_{\tau,\mathcal{AT}}^w(\Lambda) := \{\mathbf{v} = (\mathbf{v}_0, \ldots, \mathbf{v}_{m-1}) \in \ell_2(\Lambda),\ \mathbf{v}_i \in \ell_{\tau,\mathcal{T}}^w(\{i\} \times \Lambda^\square)\}$$

with the natural quasi-seminorm

$$|\mathbf{v}|_{\ell_{\tau,\mathcal{AT}}^w(\Lambda)} := \max_{i=0}^{m-1} |\mathbf{v}_i|_{\ell_{\tau,\mathcal{T}}^w(\Lambda^\square)}, \quad \mathbf{v} = (\mathbf{v}_0, \ldots, \mathbf{v}_{m-1}).$$

Because $\ell_{\tau,\mathcal{T}}^w(\Lambda^\square)$ is contained in $\mathcal{A}_{\mathcal{T}}^s$, $\frac{1}{\tau} = s + \frac{1}{2}$, we also have $\ell_{\tau,\mathcal{AT}}^w(\Lambda) \subset \mathcal{A}_{\mathcal{AT}}^s$, whereas the converse inclusion does not hold in general. Similarly to the case of unrestricted N-term approximation, the results from approximation with wavelet bases almost directly carry over to the case of aggregated wavelet frames if we have at hand a smooth partition of unity.

Proposition 3.4.7 ([76, Prop. 3.46]). *Assume that there exists a smooth partition of unity $\{\sigma_i\}_{i=0}^{m-1}$ with respect to the overlapping domain decomposition $\{\Omega_i\}_{i=0}^{m-1}$ of Ω. Let $v \in B_{\bar{\tau},\bar{\tau}}^{sd+t}(\partial\Omega, t)$, $0 < s < \frac{l-t}{d}$, $\frac{1}{\bar{\tau}} < \frac{1}{\tau} = s + \frac{1}{2}$ and $sd + t - \frac{1}{\tau} \notin \{0, \ldots, l-2\}$. Then, the expansion coefficients $\mathbf{c} \in \ell_2(\Lambda)$, $\mathbf{c}^\top \Psi^{H_0^t(\Omega)} = v$, calculated with the help of the scaled non-canonical dual L_2-frame $\mathbf{D}^t \tilde{\Psi}^{L_2(\Omega)}$ from Proposition 2.2.16,*

$$c_\lambda = \langle v, 2^{t|\lambda|} \sigma_i \tilde{\psi}_{(i,\lambda)} \rangle, \quad \lambda \in \{i\} \times \Lambda_i, \tag{3.10}$$

are contained in $\mathcal{A}_{\mathcal{AT}}^s$.

For the L-shaped domain, this result again does not apply. Fortunately, under slightly stronger regularity assumptions, a similar characterization can be shown even in this case.

Proposition 3.4.8 ([76, Prop. 3.47]). *Let the assumptions from Proposition 3.3.7 be fulfilled. Then, the expansion coefficients $\mathbf{c} = (c_\lambda)_{\lambda \in \Lambda}$ defined analogously to Equation (3.10) are contained in $\mathcal{A}_{\mathcal{AT}}^{\tilde{s}}$ for all $0 < \tilde{s} < \breve{s}$ with $0 < \breve{s} < \min\{\frac{l-t}{2}, \frac{\eta^* s + t - 1}{2(t+\eta^*-1)} - \frac{1}{2}\}$, where η^* was defined in (3.9).*

To sum up, we conclude that even for the L-shaped domain, the additional restriction to aggregated tree structures only has an arbitrarily small effect on the expected convergence rate of the best N-term approximation. In fact, the proofs in [76] show even a slightly stronger result.

Remark 3.4.9. *Proposition 3.4.7 and Proposition 3.4.8 remain true if $\mathcal{A}_{\mathcal{AT}}^s$ and $\mathcal{A}_{\mathcal{AT}}^{\tilde{s}}$ are replaced by $\ell_{\tau,\mathcal{AT}}^w(\Lambda)$, $\frac{1}{\tau} = s + \frac{1}{2}$ and $\ell_{\tilde{\tau},\mathcal{AT}}^w(\Lambda)$, $\frac{1}{\tilde{\tau}} = \tilde{s} + \frac{1}{2}$, respectively.*

3.4.4 Aggregated tree approximation and asymptotic optimality

We have stated before that the rate of the best N-term (tree) approximation will serve as a benchmark for our numerical algorithms. With the tools and concepts introduced in the previous subsections, we can now define this benchmark more explicitly. Recall that $\mathbf{u} \in \mathcal{A}_{\mathcal{AT}}^s$ implies that for each $N \in \mathbb{N}$, there exists a tree-structured vector $\mathbf{u}_N \in \ell_2(\Lambda)$ with $\#\operatorname{supp} \mathbf{u}_N \leq N$ such that

$$\|\mathbf{u} - \mathbf{u}_N\|_{\ell_2(\Lambda)} \leq (N+1)^{-s} \|\mathbf{u}\|_{\mathcal{A}_{\mathcal{AT}}^s}.$$

Conversely, if we fix $\varepsilon > 0$ and set $\mathbf{u}_\varepsilon := \mathbf{u}_N$ with $N \in \mathbb{N}$ minimal such that

$$(N+1)^{-s} \|\mathbf{u}\|_{\mathcal{A}_{\mathcal{AT}}^s} \leq \varepsilon,$$

then we have

$$\|\mathbf{u}_\varepsilon^\top \Psi - \mathbf{u}^\top \Psi\|_{H^t(\Omega)} \lesssim \|\mathbf{u}_\varepsilon - \mathbf{u}\|_{\ell_2(\Lambda)} \leq \varepsilon$$

and

$$N + 1 \geq \varepsilon^{-1/s} \|\mathbf{u}\|_{\mathcal{A}_{\mathcal{AT}}^s}^{1/s}.$$

This is essentially the same as before, but seen from a different point of view. We now fix a tolerance and are interested in the degrees of freedom are needed to achieve this tolerance. We see from the above arguments that this quantity is of the order $\varepsilon^{-1/s} \|\mathbf{u}\|_{\mathcal{A}_{\mathcal{AT}}^s}^{1/s}$, which motivates the following definition.

Definition 3.4.10. *A numerical algorithm for the approximation of a possibly unknown $u = \mathbf{u}^\top \Psi^{H_0^t(\Omega)} \in H_0^t(\Omega)$ is called* asymptotically optimal *if it has the following properties:*

- *For any given tolerance $\varepsilon > 0$, the algorithm terminates with an approximation $\mathbf{u}_\varepsilon \in \ell_2(\Lambda)$ such that $\|u - \mathbf{u}_\varepsilon^\top \Psi\|_{H^t(\Omega)} \lesssim \varepsilon$.*

- *There exists a number $s^* > 0$ such that $\mathbf{u} \in \mathcal{A}_{\mathcal{AT}}^s$, $0 < s < s^*$ implies $\#\operatorname{supp} \mathbf{u}_\varepsilon \lesssim \varepsilon^{-1/s} + 1$ with a constant possibly depending on $\mathbf{u}$, but not on ε. Moreover, an analogous estimate is required for the number of operations needed to calculate $\mathbf{u}_\varepsilon$.*

Remark 3.4.11. *The additive constant we allow for in the support and complexity estimates in the above definition is of minor importance, because we are mostly interested in estimates for the case $\varepsilon \to 0$, which means that $\varepsilon^{-1/s}$ becomes large. In practical implementations of a method, an additive constant in the computational time will occur in any case, since some work is needed for setting up the problem or doing precomputations.*

3.5 Regularity of solutions to elliptic equations

We have seen in the previous sections that the convergence rate of nonlinear approximation in $H^t(\Omega)$ or in $\ell_2(\Lambda)$ for the corresponding discretized problem can be linked to the Besov regularity of the function to approximate in the adaptivity scale $B^{sd+t}_{\tau,\tau}(\Omega)$, $\frac{1}{\tau} = s + \frac{1}{2}$ or in a very similar scale as in the case of aggregated frames on the L-shaped domain. The convergence rate of classical linear iteration schemes such as the Galerkin method in the spaces V_j was described in terms of classical Sobolev regularity in the scale $H^{sd+t}(\Omega)$. We will see that in many cases the Besov regularity of the solution in the adaptivity scale is higher than its Sobolev regularity. Hence, in these cases numerical methods for elliptic PDEs based on nonlinear approximation schemes potentially outperform standard linear schemes. For further reading on the Sobolev regularity, we refer, e.g., to [66, 67, 73]. Results on Besov smoothness can for instance be found in [30, 32, 39]. Quite recently, results of the latter kind have also been proven for further types of equations such as stochastic PDEs or the Stokes- and Navier-Stokes systems, see, e.g., [22] and [61], respectively.

3.5.1 Regularity of linear elliptic equations

At first, let us recall a few important results for the case of *linear* elliptic PDEs. As a standard reference problem, most results deal with the weak form of the Poisson equation

$$-\Delta u = f \text{ in } \Omega, \quad u = 0 \text{ on } \partial\Omega. \tag{3.11}$$

On smooth domains, the Sobolev regularity of the solution is imposed by the Sobolev regularity of the right-hand side and we gain the maximum shift that we can expect. This is stated in the following result, which is a direct application of [68, Th. 9.1.16].

Proposition 3.5.1. *Let $\Omega \subset \mathbb{R}^d$ be a C^k-domain, $k \geq 1$ and $f \in H^\vartheta(\Omega)$, where $-1 \leq \vartheta < k-2$ and $\vartheta \neq -\frac{1}{2}$. Then, the Poisson equation has a unique weak solution $u \in H_0^1(\Omega) \cap H^{\vartheta+2}(\Omega)$.*

In many applications, however, the domain Ω is not as smooth as required by Proposition 3.5.1. For instance, our prototype of a polygonal domain in $\mathbb{R}^2$, the L-shaped domain, is only a Lipschitz domain. In such situations, Sobolev regularity of the solution can only be shown within a limited range of parameters, see the following result.

Theorem 3.5.2 ([73, Th. B.2]). *Let Ω be a bounded Lipschitz domain in $\mathbb{R}^d$, $d \geq 2$ and $f \in L_2(\Omega)$. Then, the Poisson equation (3.11) has a unique weak solution $u \in H_0^1(\Omega) \cap H^{3/2}(\Omega)$.*

This result is sharp in the sense that for general Lipschitz domains, a Sobolev regularity higher than $\frac{3}{2}$ cannot be expected even if the right-hand side is arbitrarily smooth.

Proposition 3.5.3 ([73, Th. A.3]). *For each $\vartheta > \frac{3}{2}$, there exist a Lipschitz domain Ω and a right-hand side $f \in C^\infty(\overline{\Omega})$ such that the solution u to (3.11) is not contained in $H^\vartheta(\Omega)$.*

Now let us have a look at regularity in the Besov scale. We first give a result that is valid for general Lipschitz domains. This is a special case of [32, Th. 4.1].

Proposition 3.5.4. *Let $\Omega \subset \mathbb{R}^d$ be a bounded Lipschitz domain and let $f \in H^\vartheta(\Omega)$, $\vartheta \geq -\frac{1}{2}$. Then, the solution u to (3.11) is contained in $B^\alpha_{\tau,\tau}(\Omega)$ for all $0 < \alpha < \min\{\vartheta + 2, 1 + \frac{1}{\tau}\}$ with $\frac{d-1}{d+1} < \tau \leq 2$.*

Let us illustrate and discuss this result for the case $d = 2$. If $f \in L_2(\Omega)$, then the proposition above states that the solution u is contained in all spaces $B^\alpha_{\tau,\tau}(\Omega)$, where $0 < \alpha < \min\{2, 1 + \frac{1}{\tau}\}$ and $\frac{1}{2} \leq \frac{1}{\tau} < 3$. In Figure 3.4, the boundary of this area is sketched with the solid lines. The dashed line is the embedding line into $H^1(\Omega)$, which exactly corresponds to the scale of Besov spaces $B^{sd+t}_{\tau,\tau}(\Omega) = B^{2s+1}_{\tau,\tau}(\Omega)$, $\frac{1}{\tau} = s + \frac{1}{2}$ we are interested in to describe the approximation rate of nonlinear schemes in $H^1(\Omega)$. Calculating the intersection of this line with the boundary of the area that indicates the Besov regularity of u, we see that the solution is contained in all the spaces $B^{2s+1}_{\tau,\tau}(\Omega)$, $\frac{1}{\tau} = s + \frac{1}{2}$ with $0 \leq s < \frac{1}{2}$. Also note that increasing the Sobolev regularity of the right-hand side beyond $f \in L_2(\Omega)$ does not guarantee a higher Besov regularity of the solution in the relevant scale. Nevertheless, comparing this result to the $H^{3/2}$-result in Theorem 3.5.2, we see that in the scale $H^{sd+t}(\Omega) = H^{2s+1}(\Omega)$, we only get up to $s = \frac{1}{4}$. Hence, in view of the results from the previous sections, the convergence rate of the best N-term wavelet approximation is nearly twice as high compared to the linear approximation scheme based on approximation in the multiresolution spaces V_j with increasing j.

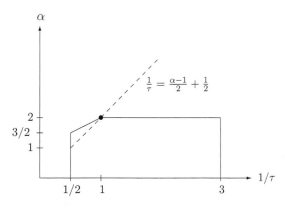

Figure 3.4: A DeVore-Triebel diagram illustrating the result of Proposition 3.5.4.

This result for general bounded Lipschitz domains already makes a strong case for further investigating nonlinear approximation schemes. In many applications and examples, however, we know more about the domain than only that it is Lipschitz. Often, one has to deal with polygonal domains, i.e., domains whose boundary is a polygon, such as the L-shaped domain. These domains are not smooth, and the solution of the Poisson equation may have singularities no matter how smooth the right-hand side is. However, it seems plausible that for such domains, a higher Besov regularity than on general Lipschitz domains can be shown. This is because on polygonal domains, the singularities caused by their geometry occur as point singularities in the relatively few corners. On general Lipschitz domains, however, all boundary points could be problematic.

To better explain the situation on polygonal domains, it is vital that the singularities caused by the geometry of such domains can be described explicitly with the help of the results from Sections 2.4 and 2.7 in [67], see also [30]. There, it is shown that for $f \in H^\vartheta(\Omega)$, $\vartheta \geq -1$, the solution of the Poisson equation on a polygonal domain in $\mathbb{R}^2$ with interior angles ω_l, $l = 1, \ldots, N$ at the corners S_l can be written as the sum of a regular part and a singular part,

$$u = u_R + u_S,$$

where $u_R \in H^{\vartheta+2}(\Omega)$, which means that this term has the maximum expected regularity driven only by the right-hand side. To describe the latter term, define the *singularity functions*

$$S_{l,k} := \eta_l(r_l) r_l^{\lambda_{l,k}} \sin(\theta_l \lambda_{l,k}), \quad \lambda_{l,k} := \frac{k\pi}{\omega_l} \notin \mathbb{Z}, \tag{3.12}$$

$$S_{l,k} := \eta_l(r_l) r_l^{\lambda_{l,k}} \left(\log r_l \sin(\theta_l \lambda_{l,k}) + \theta_l \cos(\theta_l \lambda_{l,k}) \right), \text{ otherwise} \tag{3.13}$$

written in polar coordinates (θ_l, r_l) with respect to the corner S_l. Here, $\eta_l \in C^\infty(\mathbb{R}^+)$ denotes a truncation function, i.e., it is $\eta_l(0) = 1$ and $\eta_l(x) = 0$ for large x so that the functions $S_{l,k}$ do not interfere with singularity functions stemming from other corners. Assuming that $\lambda_{l,k} \neq s + 1$ for each l, k, the singular part of the solution can then be written as a linear combination of the singularity functions,

$$u_S = \sum_{l=0}^{N} \sum_{0 < \lambda_{l,k} < \vartheta+1} c_{l,k} S_{l,k}.$$

For the case $\vartheta = 0$, i.e., a right-hand side in $L_2(\Omega)$, this simplifies to

$$u_S = \sum_{l=0}^{N} \sum_{\omega_l > \pi} c_{l,1} S_{l,1}.$$

and only Formula (3.12) for the singularity function applies. From [67, Th. 1.2.18], it follows that $S_{l,1} \in H^\vartheta(\Omega)$ if and only if $\vartheta < 1 + \frac{\pi}{\omega_l}$. Hence, for non-convex polygonal domains, the Sobolev regularity of the singular part is less than 2. For instance, on the L-shaped domain with a maximum interior angle $\omega = \frac{3}{2}\pi$, the singular part, and hence also the solution u, is only contained in $H^\vartheta(\Omega)$ for $\vartheta < \frac{5}{3}$. In the scale $H^{2s+1}(\Omega)$ governing the convergence of the standard linear approximation scheme, this corresponds to a rate $s < \frac{1}{3}$. Regarding the Besov regularity of the solution, however, we have the following, much stronger result. This result is also obtained by studying regularity of the singular part and by using interpolation.

Theorem 3.5.5 ([30, Th. 2.4]). *Suppose that $f \in H^\vartheta(\Omega)$, $\vartheta > -\frac{1}{2}$ and $\frac{(\vartheta+1)\omega_l}{\pi} \notin \mathbb{N}$ for each $l = 0, \ldots, N$. Then, the solution u to the Poisson equation is contained in $B^{2s+3/2}_{\tau,\tau}(\Omega)$ for all $0 < s < \frac{\vartheta}{2} + \frac{1}{4}$, where $\frac{1}{\tau} = s + \frac{1}{2}$.*

This result shows that, contrary to classical Sobolev regularity, the Besov regularity in the scale $B^{2s+1}_{\tau,\tau}(\Omega)$, $\frac{1}{\tau} = s + \frac{1}{2}$ is only governed by the regularity of the right-hand side. Hence, given sufficient smoothness of the right-hand side, and in view of, for instance, Proposition 3.3.2, the rate of the best N-term approximation is only limited by the approximation power of the underlying wavelet basis.

3.5.2 Regularity of nonlinear elliptic equations

Let us now consider a model problem for semilinear equations, namely the problem

$$-\Delta u(x) + G(u(x)) = f(x), \quad x \in \Omega \tag{3.14}$$

understood in the weak form with Dirichlet boundary conditions, where $\Omega \subset \mathbb{R}^d$ is a bounded Lipschitz domain and $G : \mathbb{R} \to \mathbb{R}$ a function that induces a Nemitsky operator $\mathcal{G} : H^1_0(\Omega) \to H^{-1}(\Omega)$, see also Section 1.2. Note that this equation is a special case of (1.10), but a representative example of order two, because the Laplace operator is the prototype of an elliptic operator of this order. For equations of this form and under additional assumptions on the nonlinear part, Sobolev regularity for the solution of this problem can be shown, but only in a limited range. This result is a direct consequence of [39, Th. 3].

Proposition 3.5.6. *Assume that G is continuous and that there exist $a, b \in \mathbb{R}$ such that*

$$|G(v)| \le a + b|v|, \quad v \in \mathbb{R}. \tag{3.15}$$

If b is sufficiently small and $f \in H^\vartheta(\Omega)$, $-1 \le \vartheta \le -\frac{1}{2}$, then Problem 3.14 has a weak solution $u \in H^1_0(\Omega) \cap H^\alpha(\Omega)$, $\alpha < \vartheta + 2$.

Hence, even for this kind of semilinear equations, the Sobolev regularity is shifted by almost two compared to the regularity of the right-hand side. However, this result is only valid in a limited range of parameters, a Sobolev regularity of $\frac{3}{2}$ or higher cannot be expected even if the right-hand side is arbitrarily smooth. Recall that this estimate cannot be significantly sharpened in the sense that a higher Sobolev regularity than

$\frac{3}{2}$ is not given in general even for linear problems, see Subsection 3.5.1. The reason for this lies in singularities that can be induced by the geometry of the domain. In Proposition 3.5.4, we have seen that, for linear problems, a higher Besov regularity could be verified compared to classical Sobolev regularity. A similar result also holds for the semilinear problems we deal with in this thesis. This is a special case of [39, Th. 7].

Theorem 3.5.7. *Let Ω be a bounded Lipschitz domain in $\mathbb{R}^d$, $d = 2, 3$. Assume that the nonlinearity fulfills an estimate*

$$|G(x) - G(y)| \leq c|x - y|, \quad x, y \in \mathbb{R}$$

with a sufficiently small constant $c \in \mathbb{R}$. Let $f \in H^\vartheta(\Omega)$, $\vartheta > 0$. Then, the perturbed Poisson equation (3.14) has a solution $u \in B^\alpha_{1/2,1/2}(\Omega)$ for all $\alpha < 3$.

Although this result does not directly provide us with a regularity result in the spaces $B^{sd+t}_{\tau,\tau}(\Omega)$, $\frac{1}{\tau} = s + \frac{1}{2}$, we can combine it with the Sobolev regularity from Propositon 3.5.6 and use interpolation of Besov spaces to end up in the desired scale. For simplicity, we restrict ourselves to the case $d = 2$.

Corollary 3.5.8. *Let the conditions from Theorem 3.5.7 be fulfilled and $d = 2$. Then, the solution u is contained in the Besov spaces $B^{sd+t}_{\tau,\tau}(\Omega) = B^{2s+1}_{\tau,\tau}(\Omega)$, $\frac{1}{\tau} = s + \frac{1}{2}$ for all $0 \leq s < \frac{1}{2}$.*

Proof. Estimate (3.15) holds with $a := |G(0)|$ and $b := c$. Hence, we can apply Proposition 3.5.6 and obtain $u \in H^{3/2-\varepsilon}(\Omega) = B^{3/2-\varepsilon}_{2,2}(\Omega)$ for an arbitrarily small $\varepsilon > 0$. By Theorem 3.5.7, the solution is also contained in $B^{3-\varepsilon}_{1/2,1/2}(\Omega)$. Using Proposition 3.1.2 with $\theta = \frac{2}{3} + \frac{2}{3}\varepsilon$, we see that $u \in B^{2s+1}_{\tau,\tau}(\Omega)$, where $s = \frac{1}{2} - \varepsilon$ and $\frac{1}{\tau} = s + \frac{1}{2}$ as illustrated in Figure 3.5. This shows the assertion. □

We have seen that for linear elliptic problems, stronger regularity results can be shown if we restrict ourselves to polygonal domains. At least under a few additional assumptions on the nonlinear part, in this case some improvements can also be derived for semilinear equations. To formulate the necessary assumptions, we denote by $U^1_2(\mathbb{R})$ the set of all differentiable Lipschitz continuous functions $G : \mathbb{R} \to \mathbb{R}$ such that

$$|G'|_{U_2(\mathbb{R})} := \sup_{\rho > 0} \rho^{-1/2} \left(\int_{\mathbb{R}} \sup_{|h| \leq \rho} |G'(x + h) - G'(x)|^2 \, dx \right)^{1/2}$$

is finite.

Theorem 3.5.9 ([39, Th. 8]). *Let Ω be a simply connected polygonal domain with interior angles ω_l, $l = 0, \ldots, N$. Let $f \in H^\vartheta(\Omega)$ for some $-\frac{1}{2} < \vartheta < \frac{3}{2}$. Assume in addition that*

$$\frac{(\vartheta + 1)\omega_l}{\pi} \notin \mathbb{N}, \quad l = 0, \ldots, N$$

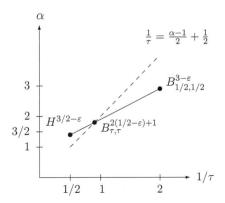

Figure 3.5: A DeVore-Triebel diagram illustrating the proof of Corollary 3.5.8.

and that $G \in U_2^1(\mathbb{R})$ with

$$\|G'\|_{L_\infty(\mathbb{R})} + |G'|_{U_2(\mathbb{R})}$$

sufficiently small. Then, the problem (3.14) has a solution $u \in B_{\tau,\tau}^\alpha(\Omega)$, $\frac{1}{\tau} = \frac{\alpha}{2} + \frac{1}{2}$ for all $\alpha < \vartheta + 2$.

This result does not yet provide us with Besov regularity in the scale $B_{\tau,\tau}^{sd+t}(\Omega)$. However, similarly to the previous subsection, this can be achieved using the Sobolev regularity and interpolation, see also [39, Section 6] for further results of this kind.

Proposition 3.5.10. *Let the conditions from Theorem 3.5.9 be fulfilled. Then, Equation (3.14) has a solution $u \in B_{\tau,\tau}^{2s+1}(\Omega)$, $\frac{1}{\tau} = s + \frac{1}{2}$ for all $s < \frac{1}{6}(\vartheta + 2)$.*

Proof. If $\|G'\|_{L_\infty(\mathbb{R})}$ is small, the conditions from Proposition 3.5.6 are fulfilled as well with $a := |G(0)|$ and $b := \|G'\|_{L_\infty(\mathbb{R})}$. Hence, we have $u \in H^{3/2-\varepsilon}(\Omega) = B_{2,2}^{3/2-\varepsilon}(\Omega)$ for arbitrarily small $\varepsilon > 0$. On the other hand, by Theorem 3.5.9, u is contained in the spaces $B_{\tau,\tau}^{\vartheta+2-\varepsilon}(\Omega)$, $\frac{1}{\tau} = \frac{\vartheta+2-\varepsilon}{2} + \frac{1}{2}$ for $\varepsilon > 0$. Thus, by using the interpolation result from Proposition 3.1.2, we have that u is also contained in the spaces $B_{\tau,\tau}^\alpha(\Omega)$ with

$$\alpha = \theta\left(\frac{3}{2} - \varepsilon\right) + (1-\theta)(\vartheta + 2 - \varepsilon), \quad \frac{1}{\tau} = \frac{\theta}{2} + (1-\theta)\left(\frac{\vartheta + 2 - \varepsilon}{2} + \frac{1}{2}\right) \quad (3.16)$$

and $0 < \theta < 1$. By a short calculation, we see that with the choice $\theta = \frac{1}{3/2-\varepsilon}$, we end up on the H^1-embedding line $\frac{1}{\tau} = \frac{\alpha-1}{2} + \frac{1}{2}$, that corresponds to the scale of spaces $B_{\tau,\tau}^{2s+1}(\Omega)$, $\frac{1}{\tau} = s + \frac{1}{2}$, see also Figure 3.6. Reinserting this value into (3.16) gives

$$\alpha = 1 + (\vartheta + 2 - \varepsilon)\left(1 - \frac{1}{3/2 - \varepsilon}\right)$$

and τ accordingly. Thus, with $\varepsilon \downarrow 0$ and $\alpha = 2s + 1$, the assertion is shown. $\qquad \square$

If we have a relatively smooth right-hand side $f \in H^{3/2}(\Omega)$, this result shows that we have $u \in B^{2s+1}_{\tau,\tau}(\Omega)$ up to $s < \frac{7}{12}$. This is at least a slight improvement compared to general Lipschitz domains as seen in Corollary 3.5.8. Moreover, recall that already for linear problems on the L-shaped domain, a Sobolev regularity of $\frac{5}{3}$ or higher could not be expected, which gives a rate $s < \frac{1}{3}$ of linear approximation schemes. Hence, also for the case of polygonal domains, it seems worthwhile to develop and study algorithms that aim at realizing the rate of the best N-term approximation. This will be the focus of the subsequent chapters.

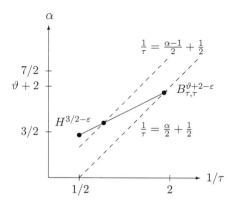

Figure 3.6: A DeVore-Triebel diagram illustrating the proof of Proposition 3.5.10.

Remark 3.5.11. *For wavelet approximation subject to tree structured index sets, such as outlined in Section 3.4, the convergence rate of the best N-term approximation was governed by the Besov regularity in a slightly different scale $B^{sd+t}_{\tilde{\tau},\tilde{\tau}}(\Omega)$, $\frac{1}{\tilde{\tau}} < \frac{1}{\tau} = s + \frac{1}{2}$. Nevertheless, the proofs and results from this section hold nearly verbatim by using interpolation with respect to the embedding line into $H^{1+\nu}(\Omega)$, with $\nu > 0$ arbitrarily small, i.e., a line slightly above the H^1-embedding line $\frac{1}{\tau} = \frac{\alpha-1}{d} + \frac{1}{2}$.*

Chapter 4

Adaptive wavelet Schwarz methods

In this chapter, the main results of this thesis are presented. Our aim is to construct adaptive wavelet methods for nonlinear equations that are not only convergent, but also optimal in the sense that they reproduce the convergence rate of the best N-term tree approximation. We will work with an overlapping domain decomposition approach that corresponds naturally to the strategy applied in Section 2.2, where we have used overlapping subdomains for the construction of an aggregated wavelet frame. The basic idea of the methods we apply is to decompose the problem into a series of subproblems on the subdomains of the domain Ω, on which the original problem is posed. This kind of domain decomposition algorithms is commonly known as overlapping *Schwarz methods*. This approach is particularly convenient, because on the subdomains, we have available Riesz bases. Hence, for the solution of the subproblems, we can in principle resort to known adaptive wavelet solvers as in [25, 26]. Moreover, working with a Riesz basis implies that, at least at the stage of the subproblems, we are not concerned with redundancies.

In this chapter, we will consider to types of Schwarz methods with specific properties: In the *multiplicative Schwarz method*, the subproblems are solved sequentially, whereas in the *additive Schwarz method*, they are independent of each other and can thus be solved in parallel. Originally, techniques of this kind have been applied to theoretically derive existence results, see [103]. In recent decades, Schwarz methods have come into focus for the numerical solution of both linear and nonlinear PDEs, mostly within the context of finite element methods, see, e.g., [58, 84, 112, 126]. They have proven to be very effective, because in many cases, the problems on the subdomains can be solved at a much lower computational cost than the problem on the entire domain because of their smaller sizes and a possibly simpler geometry. For an overview of Schwarz methods, we also refer to [96, 115]. Quite recently, in [111, 123] these methods were combined with adaptive wavelet techniques to solve linear elliptic PDEs. The convincing theoretical and numerical results lead us to exploring the application of such methods to nonlinear problems.

In a first step, following lines from [85], we present the basic ideas of the algorithms in an idealized form. To formulate an implementable version of these algorithms based on discretization with wavelets, we have to introduce a couple of elementary building blocks, which have been designed in a number of previous works. With the help of these tools, we can construct the adaptive wavelet algorithms. Finally, we shall see that these algorithms are convergent and asymptotically optimal, at least under some technical assumptions. Most results regarding the additive method have previously been published in [82].

4.1 Basic principles

Before we are ready to present the basic ideas of the Schwarz algorithms, let us shortly recall the general setting. We consider the weak form

$$\mathcal{L}u + \mathcal{G}(u) = f, \tag{4.1}$$

of a semilinear elliptic equation $Lu + G(u) = f$ with Dirichlet boundary conditions on a Lipschitz domain Ω, see Section 1.2. Since the linear operator corresponds to a bilinear form $a(\cdot, \cdot)$, we will also use the equivalent form

$$a(u, v) + \langle \mathcal{G}(u), v \rangle = \langle f, v \rangle, \quad v \in H_0^t(\Omega).$$

We assume furthermore that we have at hand an overlapping domain decomposition $\{\Omega_i\}_{i=0}^{m-1}$ of Ω into $m \geq 2$ subdomains. We will tacitly assume that this is the same decomposition that we have already used in Subsection 2.2.5, where a wavelet frame was constructed as a union of wavelet bases on the subdomains.

4.1.1 Multiplicative Schwarz method

The basic idea of the multiplicative Schwarz method is best explained for a linear problem $Lu = f$ of order two in classical form with Dirichlet boundary conditions. Starting with some initial guess $u^{(n)} = v_0^{(n)}$ for the solution, on each subdomain Ω_i, $k = 0, \ldots, m-1$, we subsequently solve the *local problems*

$$L(v_{i+1}^{(n)}) = f \quad \text{on } \Omega_i, \tag{4.2}$$

$$v_{i+1}^{(n)} = v_i^{(n)} \quad \text{on } \Omega \setminus \Omega_i, \tag{4.3}$$

$$v_{i+1}^{(n)} = 0 \quad \text{on } \partial\Omega \setminus \Omega_i. \tag{4.4}$$

Finally, we set $u^{(n+1)} := v_m^{(n)}$ to obtain the next *global iterate*. This procedure is repeated until a given stopping criterion is fulfilled. From this formulation it is easy to see that the information transmission between different subdomains takes place only via boundary conditions. Alternatively, due to the linearity of L, the local problems can be seen as the problems of finding a *local correction* $e_i^{(n)} = v_{i+1}^{(n)} - v_i^{(n)}$ from

$$L(e_i^{(n)}) = f - L(v_i^{(n)}) \quad \text{on } \Omega_i$$

with zero boundary conditions on $\partial\Omega_i$. To generalize this approach to nonlinear problems, there are two general ways of proceeding. One idea is to work directly with local nonlinear subproblems. The other approach, which we choose to follow, is to linearize the problem in each step and work with linear subproblems. This has two main advantages. Firstly, we expect that the numerical solution of linear subproblems can be performed much faster than that of nonlinear subproblems. Secondly, using linearity as above, one can reduce the subproblems to finding a local correction term $e_i^{(n)}$ with zero boundary conditions along the inner interfaces. This way, we can avoid the implementation of trace operators and use ansatz spaces with zero boundary conditions for the local subproblems. A straightforward linearization strategy, also known as *Picard iteration*, is given by shifting the nonlinearity to the right-hand side and evaluating it on the last global iterate. Hence, instead of Equation (4.2), we solve

$$L(v_{i+1}^{(n)}) = f - G(u^{(n)}) \text{ on } \Omega_i \qquad (4.5)$$

subject to the boundary conditions (4.3) and (4.4). As before, we can equivalently solve

$$L(e_i^{(n)}) = -L(v_i^{(n)}) + f - G(u^{(n)}) \text{ on } \Omega_i \qquad (4.6)$$

with zero boundary conditions for $e_i^{(n)}$, and set $v_{i+1}^{(n)} = v_i^{(n)} + e_i^{(n)}$. Combining these ideas and going back to the weak formulation and more general linear operators leads us to the following idealized algorithm, which is a modified version of a method presented in [85]. For simplicity, in this version we do not yet give an explicit stopping criterion.

Algorithm 1 MultSchw1

$u^{(0)} := 0$
for $n = 0, 1, \ldots$ **do**
$\quad v_0^{(n)} := u^{(n)}$
$\quad$ **for** $i = 0, \ldots, m-1$ **do**
$\quad\quad$ Compute $e_i^{(n)} \in H_0^t(\Omega_i)$ such that
$\quad\quad\quad a(e_i^{(n)}, v) = -a(v_i^{(n)}, v) + f(v) - \mathcal{G}(u^{(n)})(v), \quad v \in H_0^t(\Omega_i).$
$\quad\quad v_{i+1}^{(n)} := v_i^{(n)} + e_i^{(n)}$
$\quad$ **end for**
$\quad u^{(n+1)} := v_m^{(n)}$
end for

As a first step towards a convergent and asymptotically optimal adaptive wavelet method, we show the convergence of this idealized algorithm. To do so, we have to fix some notation. From now on, let P_i denote the $a(\cdot, \cdot)$-orthogonal projector onto $H_0^t(\Omega_i)$, which is defined by

$$a(v, w) = a(P_i v, w), \quad v \in H_0^t(\Omega), \ w \in H_0^t(\Omega_i). \qquad (4.7)$$

These orthogonal projectors play a major role in the convergence theory of Schwarz methods. For further reading, we refer to standard literature on functional analysis such as [79, Section 9.5]. Using these projectors, we set

$$\theta := \|\!|(I - P_{m-1}) \cdot \ldots \cdot (I - P_0)|\!\|\tag{4.8}$$

measured in the operator norm $\|\!|\cdot|\!\|$ induced by the energy norm $\|\!|\cdot|\!\| = a(\cdot,\cdot)^{1/2}$. With these notations at hand, we can formulate the following convergence result, which is based on [85, Theorem 1].

Proposition 4.1.1. *Let the assumption from Lemma 1.0.1 be fulfilled. Then, for the iterates $u^{(n)}$ from Algorithm 1, it holds that*

$$\|\!|u^{(n)} - u|\!\| \leq \rho^n \|\!|u^{(0)} - u|\!\|,$$

where $\rho := \theta + mc$.

Proof. By definition of $e_i^{(n)}$, and using that $a(u,v) + \mathcal{G}(u)(v) = f(v)$ for all $v \in H_0^t(\Omega_i)$, we have

$$
\begin{aligned}
a(e_i^{(n)}, v) &= -\mathcal{G}(u^{(n)})(v) + f(v) - a(v_i^{(n)}, v) \\
&= -\mathcal{G}(u)(v) + f(v) + (\mathcal{G}(u)(v) - \mathcal{G}(u^{(n)})(v)) - a(v_i^{(n)}, v) \\
&= a(u, v) + (\mathcal{G}(u)(v) - \mathcal{G}(u^{(n)})(v)) - a(v_i^{(n)}, v) \\
&= a(u - v_i^{(n)}, v) - (\mathcal{G}(u^{(n)})(v) - \mathcal{G}(u)(v)) \\
&= a(P_i(u - v_i^{(n)}), v) - (\mathcal{G}(u^{(n)})(v) - \mathcal{G}(u)(v)).
\end{aligned}
$$

Now consider the linear operator on the subdomains

$$\mathcal{L}_i : H_0^t(\Omega_i) \to H^{-t}(\Omega_i), \ v \mapsto \mathcal{L}_i v := a(v, \cdot).$$

Because the bilinear form is elliptic on $H_0^t(\Omega_i) \subset H_0^t(\Omega)$, this operator is boundedly invertible. Hence, we can rewrite the above equation as

$$e_i^{(n)} = P_i(u - v_i^{(n)}) - \mathcal{L}_i^{-1}(\mathcal{G}(u^{(n)}) - \mathcal{G}(u)).\tag{4.9}$$

Therefore, we have

$$v_{i+1}^{(n)} - u = e_i^{(n)} + v_i^{(n)} - u = (I - P_i)(v_i^{(n)} - u) - \mathcal{L}_i^{-1}(\mathcal{G}(u^{(n)}) - \mathcal{G}(u))\tag{4.10}$$

and, inductively,

$$
\begin{aligned}
u^{(n+1)} - u &= v_m^{(n)} - u \\
&= (I - P_{m-1}) \cdot \ldots \cdot (I - P_0)(v_0^{(n)} - u) \\
&\quad - \sum_{i=0}^{m-1}(I - P_{m-1}) \cdot \ldots \cdot (I - P_{i+1})\mathcal{L}_i^{-1}(\mathcal{G}(u^{(n)}) - \mathcal{G}(u)).
\end{aligned}\tag{4.11}
$$

With the help of the Cauchy-Schwarz inequality, the short calculation

$$\|\mathcal{L}_i v\| = \sup_{w \in H_0^t(\Omega_i)} \frac{|a(v,w)|}{\|w\|} = \sup_{w \in H_0^t(\Omega_i)} \frac{|a(v,w)|}{a(w,w)^{1/2}} = a(v,v)^{1/2} = \|v\|$$

shows that $\mathcal{L}_i$ is an isometry with respect to the energy norm. Moreover, because the P_i are $a(\cdot,\cdot)$ orthogonal projectors onto $H_0^t(\Omega_i)$, it is $\|P_i\| = 1$ in the induced operator norm. Similarly, $(I - P_i)$ is the orthogonal projector on the complement space $H_0^t(\Omega_i)^\perp$, thus $\|I - P_i\| = 1$. Hence, from Equation (4.11), we obtain

$$\|u^{(n+1)} - u\| \le \theta \|u^{(n)} - u\| + m\|\mathcal{G}(u^{(n)}) - \mathcal{G}(u)\| \le (\theta + mc)\|u^{(n)} - u\|.$$

Using induction over n, this shows the assertion. $\qquad\square$

Thus, in order to theoretically ensure that Algorithm 1 is convergent, we have to show that the parameter θ from Equation (4.8) is smaller than one and we have to assume that the constant c is sufficiently small. For the first task, we recall the following result, which is part of [84, Theorem I.2]. Due to its importance for the remainder of this chapter, we give this result and recall its proof.

Proposition 4.1.2. *Let $\Omega = \bigcup_{i=0}^{m-1} \Omega_i$ be an overlapping domain decomposition. Then, it holds that*

$$\theta = \|(I - P_{m-1}) \cdot \ldots \cdot (I - P_0)\| < 1.$$

Proof. First of all, because $(I - P_i)$ is the $a(\cdot,\cdot)$-orthogonal projector on $H_0^t(\Omega_i)^\perp$, we have

$$\theta \le \prod_{i=0}^{m-1} \|(I - P_i)\| \le 1.$$

Since $\Omega = \bigcup_{i=0}^{m-1} \Omega_i$ is an overlapping domain decomposition and $\|\cdot\|$ is equivalent to the standard H^t-norm on $H_0^t(\Omega)$, each $v \in H_0^t(\Omega)$ can be written as $v = \sum_{i=0}^{m-1} v_i$ with $v_i \in H_0^t(\Omega_i)$ and $\left(\sum_{i=0}^{m-1} \|v_i\|^2\right)^{1/2} \lesssim \|v\|$, see Lemma 2.2.10. Using the Cauchy-Schwarz inequality twice and the properties of the projectors P_i, we obtain

$$\|v\|^2 = a(v,v) = \sum_{i=0}^{m-1} a(v,v_i) = \sum_{i=0}^{m-1} a(P_i v, v_i) \le \sum_{i=0}^{m-1} \|P_i v\| \|v_i\|$$

$$\le \left(\sum_{i=0}^{m-1} \|P_i v\|^2\right)^{1/2} \left(\sum_{i=0}^{m-1} \|v_i\|^2\right)^{1/2} \tag{4.12}$$

$$\lesssim \left(\sum_{i=0}^{m-1} \|P_i v\|^2\right)^{1/2} \|v\|,$$

from which it follows that

$$\|v\| \lesssim \left(\sum_{i=0}^{m-1} \|P_i v\|^2\right)^{1/2}. \tag{4.13}$$

Assume now that $\theta = 1$. Then, there exists a sequence $(v_n)_{n \in \mathbb{N}} \subset H_0^t(\Omega)$ with $\|v_n\| = 1$ such that $\lim_{n \to \infty} \|(I - P_{m-1}) \cdot \ldots \cdot (I - P_0)v_n\| = 1$. Because of $\|(I - P_i)\| = 1$, we moreover have

$$\lim_{n \to \infty} \|(I - P_j) \cdot \ldots \cdot (I - P_0)v_n\| = 1, \quad 0 \leq j \leq m - 1.$$

By Pythagoras' theorem, it holds that $\|w\|^2 = \|P_j w\|^2 + \|(I - P_j)w\|^2$ for each $w \in H_0^t(\Omega)$. Thus, we have

$$\lim_{n \to \infty} \|P_j(I - P_{j-1}) \cdot \ldots \cdot (I - P_0)v_n\| = 0, \tag{4.14}$$

and, in particular, $\lim_{n \to \infty} \|P_0 v_n\| = 0$. Now, by induction over j, we show that $\lim_{n \to \infty} \|P_j v_n\| = 0$ for all $0 \leq j \leq m - 1$. Hence, assume that $\lim_{n \to \infty} \|P_i v_n\| = 0$ for $0 \leq i \leq j - 1$. Because the projectors P_i are continuous, from Equation (4.14) and the induction hypothesis, we have

$$\begin{aligned}
0 &= \lim_{n \to \infty} \|P_j(I - P_{j-1}) \cdot \ldots \cdot (I - P_1)(I - P_0)v_n\| \\
&= \lim_{n \to \infty} \|P_j(I - P_{j-1}) \cdot \ldots \cdot (I - P_1)v_n\| \\
&= \ldots = \lim_{n \to \infty} \|P_j v_n\|.
\end{aligned}$$

This, however, is a contradiction to the estimate (4.13). Hence, it follows that $\theta < 1$. $\qquad \square$

Remark 4.1.3. *The parameter θ describes the error reduction in each step of the multiplicative Schwarz algorithm applied to linear problems, which correspond to the case $c = 0$. The theory from [126, Chapter 4] allows a more detailled analysis of this quantity. In particular, bounds for θ taking values significantly smaller than one can be derived in many cases. Then, there is at least a considerable range of values the constant c may take so that $\rho = \theta + mc$ remains smaller than one, i.e., the convergence of Algorithm 1 is ensured.*

With the help of Propositions 4.1.1 and 4.1.2, we can now give a sufficient criterion for the convergence of Algorithm 1.

Proposition 4.1.4. *Let $G \in C^1(\mathbb{R})$ such that $\|G'\|_\infty$ is finite and sufficiently small. Then, it is $\rho < 1$, thus Algorithm 1 is geometrically convergent.*

Proof. Using that the dual pairing on $H_0^t(\Omega) \times H^{-t}(\Omega)$ is the continuous extension of the scalar product on $L_2(\Omega)$, we obtain that

$$
\|\mathcal{G}(u) - \mathcal{G}(v)\| = \sup_{w \in H_0^t(\Omega)} \frac{|(\mathcal{G}(u) - \mathcal{G}(v))(w)|}{\|w\|}
$$

$$
= \sup_{w \in H_0^t(\Omega)} \frac{|\int_\Omega (G(u(x)) - G(v(x)))w(x)\,\mathrm{d}x|}{\|w\|}
$$

$$
\leq \sup_{w \in H_0^t(\Omega)} \sup_{z \in \mathbb{R}} |G'(z)| \|u - v\|_{L_2(\Omega)} \frac{\|w\|_{L_2(\Omega)}}{\|w\|}
$$

$$
\lesssim \|u - v\|_{L_2(\Omega)} \cdot \sup_{z \in \mathbb{R}} |G'(z)|
$$

$$
\lesssim \|u - v\| \cdot \sup_{z \in \mathbb{R}} |G'(z)|.
$$

Hence, if $\|G'\|_\infty$ is sufficiently small, so is the constant c in Lemma 1.0.1. Then, by Proposition 4.1.2, we have $\rho = \theta + mc < 1$. $\qquad\square$

So far, we have analyzed the convergence of the idealized Algorithm 1. However, in its current form, the algorithm is not yet implementable. First of all, we have not yet specified how to deal with the subproblems. Secondly, in practice we cannot expect to solve them exactly, but only up to a given tolerance. Thus, we divide the formulation of an implementable version into two steps. As a first step, we define a modified version of the algorithm, that is convergent under the same assumptions as Algorithm 1, but allows for an inexact solution of the subproblems. To do this, we aim at a somewhat smaller error reduction in each outer iteration and set

$$
\tilde{\rho} := \frac{1}{2}(1 + \rho),
$$

which is still smaller than one if we assume that ρ is smaller than one. The second step, which is the precise description of the approximate solution of the subproblems, will be postponed to Section 4.3. There, known adaptive wavelet algorithms, see, e.g., [25, 26], will be adapted to fulfill this task and give us complexity estimates for the overall method. At first, however, we now define the following modified version of the multiplicative Schwarz algorithm.

Algorithm 2 MultSchw2

$\%$ *Let* $M \geq \|u\|$

$u^{(0)} := 0$

for $n = 0, 1, \ldots$ **do**

$\quad v_0^{(n)} := u^{(n)}$

$\quad$ **for** $i = 0, \ldots, m-1$ **do**

$\quad\quad$ Compute $\tilde{e}_i^{(n)} \in H_0^t(\Omega_i)$ as an approximation to $e_i^{(n)} \in H_0^t(\Omega_i)$ from

$\quad\quad a(e_i^{(n)}, v) = -a(v_i^{(n)}, v) + f(v) - \mathcal{G}(u^{(n)})(v), \quad v \in H_0^t(\Omega_i)$

$\quad\quad$ with tolerance $\|\tilde{e}_i^{(n)} - e_i^{(n)}\| \leq \frac{M}{2m}\tilde{\rho}^n(1-\rho)$.

$\quad\quad v_{i+1}^{(n)} := v_i^{(n)} + \tilde{e}_i^{(n)}$

$\quad$ **end for**

$\quad u^{(n+1)} := v_m^{(n)}$

end for

This algorithm is convergent under the same assumptions as the exact multiplicative Schwarz algorithm, as we will see now.

Proposition 4.1.5. *Under the same assumptions as in Proposition 4.1.1, for the iterates $u^{(n)}$ from Algorithm 2, we have the error bound*

$$\|u^{(n)} - u\| \leq \tilde{\rho}^n M.$$

Proof. We use induction over n. By definition of M and $u^{(0)}$, the case $n = 0$ is clear. Assume that the assertion holds for some $n \in \mathbb{N}$. Let $\delta_i^{(n)} := \tilde{e}_i^{(n)} - e_i^{(n)}$. As in Equation (4.10), we now have

$$v_{i+1}^{(n)} - u = (I - P_i)(v_i^{(n)} - u) - \mathcal{L}_i^{-1}(\mathcal{G}(u^{(n)}) - \mathcal{G}(u)) + \delta_i^{(n)}.$$

Hence, by an inductive argument as in the proof of Proposition 4.1.1, it is

$$u^{(n+1)} - u = (I - P_{m-1}) \cdot \ldots \cdot (I - P_0)(u^{(n)} - u)$$

$$- \sum_{j=0}^{m-1} (I - P_{m-1}) \cdot \ldots \cdot (I - P_{j+1})[\mathcal{L}_j^{-1}(\mathcal{G}(u^{(n)}) - \mathcal{G}(u)) + \delta_j^{(n)}]. \quad (4.15)$$

Recall that $\|(I - P_i)\| = 1$ and that $\|\mathcal{L}_j^{-1}(\mathcal{G}(u^{(n)}) - \mathcal{G}(u))\| \leq c\|u^{(n)} - u\|$, because $\mathcal{L}_j$ is an isometry in the energy norm and by definition of c. Using these facts and the bound for $\|\delta_i^{(n)}\|$ from Algorithm 2, we obtain

$$\|u^{(n+1)} - u\| \leq \theta\|u^{(n)} - u\| + mc\|u^{(n)} - u\| + m\frac{M}{2m}\tilde{\rho}^n(1-\rho)$$

$$= \rho\|u^{(n)} - u\| + \frac{M}{2}\tilde{\rho}^n(1-\rho).$$

By the induction hypothesis, $\|u^{(n)} - u\| \leq \tilde{\rho}^n M$, we conclude

$$\|u^{(n+1)} - u\| \leq \rho\tilde{\rho}^n M + \frac{M}{2}\tilde{\rho}^n(1-\rho) = \tilde{\rho}^n M \frac{1}{2}(2\rho + (1-\rho)) = \tilde{\rho}^{n+1}M,$$

which shows the assertion. □

The next step in the development of the adaptive wavelet algorithm is the reformulation of Algorithm 2 with respect to the discretized problem. To do so and to outline the solution of the subproblems, we first have to introduce a couple of building blocks. This will be done in Section 4.2. Before this, however, let us introduce another kind of Schwarz methods, the additive Schwarz algorithm. Contrary to the multiplicative Schwarz method, this variant allows for a parallel solution of the subproblems.

4.1.2 Additive Schwarz method

Recall that the basic idea of the multiplicative Schwarz method for linear problems amounts to solving for the local correction terms $e_i^{(n)}$ from Equation (4.6). Then, the iterate $v_i^{(n)}$ is updated after each such step. An alternative introduced in [59] is to update the global iterate only after all m subproblems are solved. This has the consequence that the subproblems are independent and therefore can be parallelized. Hence, this variant has particularly gained interested due to the development of large parallel computers. The basic idea of the method is to replace Equation (4.5) by

$$L(e_i^{(n)}) = f - L(u^{(n)}) \text{ on } \Omega_i$$

with zero boundary conditions imposed on $\partial\Omega_i$. The naive approach would be to calculate the next global iterate $u^{(n+1)}$ by simply adding all local corrections to the last global iterate, $u^{(n+1)} = u^{(n)} + \sum_{i=0}^{m-1} e_i^{(n)}$. This, however, does not lead to a convergent algorithm in general, as can simply be explained. Consider the case of $m = 2$ subdomains Ω_0, Ω_1 and a solution u that is only supported in the intersection $\Omega_0 \cap \Omega_1$. Then, starting from $u^{(0)} = 0$, we calculate the local corrections $e_i^{(0)} = u_{|\Omega_i}$, ending up with $u^{(1)} = 2u$. Similarly, we have $e_i^{(1)} = -u_{|\Omega_i}$ and $u^{(2)} = 2u - 2u = 0$. Thus, the approximations $u^{(n)}$ oscillate, but do not converge. To avoid such behavior, a relaxation parameter $0 < \omega < \frac{2}{m}$ is introduced and the global update is replaced by $u^{(n+1)} = u^{(n)} + \omega \sum_{i=0}^{m-1} e_i^{(n)}$. We will see below that, after the introduction of such a relaxation parameter, convergence can be shown.

Let us now return to the more general case of semilinear problems in weak formulation. If we combine the above approach with the linearization strategy already applied for the multiplicative Schwarz algorithm, the local subproblems take the form

$$L(e_i^{(n)}) = -L(u^{(n)}) + f - G(u^{(n)}) \text{ on } \Omega_i.$$

Together with the above considerations on the necessity of a relaxation parameter and written in the weak form, this gives rise to an idealized additive Schwarz method. In similar form, this method was introduced [85]. By using this as a starting point, we will gradually develop this method into an adaptive wavelet solver and show its convergence and optimality. However, let us first present the basic algorithm.

Algorithm 3 AddSchw1

$u^{(0)} := 0$
for $n = 0, 1, \ldots$ **do**
   for $i = 0, \ldots, m-1$ **do**
      Compute $e_i^{(n)} \in H_0^t(\Omega_i)$ such that
        $a(e_i^{(n)}, v) = -a(u^{(n)}, v) + f(v) - \mathcal{G}(u^{(n)})(v), \quad v \in H_0^t(\Omega_i).$
   end for
   $u^{(n+1)} := u^{(n)} + \omega \sum_{i=0}^{m-1} e_i^{(n)}$
end for

Let us first give a convergence result for this algorithm, which is a slightly modified version of [85, Theorem 3.2], see also [82].

Proposition 4.1.6. *Let the assumption from Lemma 1.0.1 be fulfilled with a constant* $c > 0$ *and*

$$\varrho := \|I - \omega(P_0 + \ldots + P_{m-1})\| + m\omega c. \tag{4.16}$$

Then, for the iterates $u^{(n)}$ from Algorithm 3, it holds that

$$\|u^{(n)} - u\| \le \varrho^n \|u^{(0)} - u\|.$$

Proof. We use induction over n. The case $n = 0$ is obvious. Assume that the assertion holds for some $n \in \mathbb{N}$. Analogously to Equation (4.9), we now have

$$e_i^{(n)} = P_i(u - u^{(n)}) - \mathcal{L}_i^{-1}(\mathcal{G}(u^{(n)}) - \mathcal{G}(u)). \tag{4.17}$$

Hence, it follows that

$$u - u^{(n+1)} = u - u^{(n)} - \omega \sum_{i=0}^{m-1} e_i^{(n)}$$

$$= u - u^{(n)} - \omega \sum_{i=0}^{m-1} (P_i(u - u^{(n)}) - \mathcal{L}_i^{-1}(\mathcal{G}(u^{(n)}) - \mathcal{G}(u)))$$

$$= (I - \omega \sum_{i=0}^{m-1} P_i)(u - u^{(n)}) + \omega \sum_{i=0}^{m-1} \mathcal{L}_i^{-1}(\mathcal{G}(u^{(n)}) - \mathcal{G}(u))$$

Thus, using $\|\mathcal{L}_i^{-1}(\mathcal{G}(u) - \mathcal{G}(u^{(n)}))\| \le c\|u^{(n)} - u\|$ and the induction hypothesis, we conclude

$$\|u^{(n+1)} - u\| \le \varrho \|u^{(n)} - u\| \le \varrho^{n+1} \|u^{(0)} - u\|. \tag{4.18}$$

$\square$

Thus, similarly to the multiplicative Schwarz algorithm, to theoretically ensure convergence we have to verify that $\|I - \omega \sum_{i=0}^{m-1} P_i\| < 1$ and assume that c is sufficiently small. The first task will be dealt with in the well-known result below. The latter part can, for instance, be assured if $\|G'\|_\infty$ is sufficiently small, compare Proposition 4.1.4 and its proof.

Proposition 4.1.7. *Let $\Omega = \bigcup_{i=0}^{m-1} \Omega_i$ be an overlapping domain decomposition. Then, we have*

$$\|I - \omega \sum_{i=0}^{m-1} P_i\| < 1$$

for all $0 < \omega < \frac{2}{m}$.

Proof. The result can be treated within the general theory from [113], which also gives a more detailed analysis. Here, we recall a shorter proof roughly following the lines from [123, Theorem 6.5].

Consider the operator $T := \sum_{i=0}^{m-1} P_i$. Because of Equation (4.7) and the symmetry of $a(\cdot, \cdot)$, we have

$$a(P_i v, w) = a(w, P_i v) = a(P_i w, P_i v) = a(P_i v, P_i w) = a(v, P_i w), \quad v, w \in H_0^t(\Omega).$$

Hence, the P_i are symmetric, which carries over to T by linearity. Moreover, T is positive definite because of

$$a(Tv, v) = a(\sum_{i=0}^{m-1} P_i v, v) = \sum_{i=0}^{m-1} a(P_i v, v) = \sum_{i=0}^{m-1} a(P_i v, P_i v) = \sum_{i=0}^{m-1} \|P_i v\|^2 \gtrsim \|v\|^2,$$

where we used the estimate (4.13) in the last step. Hence, it holds that

$$\|I - \omega T\| = \max\{1 - \omega \lambda_{\min}(T), \, \omega \lambda_{\max}(T) - 1\}$$

with $\lambda_{\max}(T) > \lambda_{\min}(T) > 0$ being the largest and the smallest eigenvalue of T, respectively. Together with $\lambda_{\max}(T) \leq \|T\| \leq m$, this completes the proof. $\square$

Similarly to the multiplicative Schwarz algorithm, it is unrealistic to assume that the subproblems are solved exactly. Therefore, as a step towards an implementable, adaptive version of the idealized algorithm, we define a method that only requires an inexact solution of the subproblems. To do so, we leave room for small errors made in each step and aim at a smaller error reduction of

$$\tilde{\varrho} := \frac{1}{2}(1 + \varrho)$$

in each step. With these modifications, we define the following algorithm, see [82].

Algorithm 4 AddSchw2
___

> *% Let $M \geq \|u\|$.*
> $u^{(0)} := 0$
> **for** $n = 0, 1, \ldots$ **do**
>   **for** $i = 0, \ldots, m - 1$ **do**
>   Compute $\tilde{e}_i^{(n)} \in H_0^t(\Omega_i)$ as an approximation to $e_i^{(n)} \in H_0^t(\Omega_i)$ from
> $a(e_i^{(n)}, v) = -a(u^{(n)}, v) + f(v) - \mathcal{G}(u^{(n)})(v), \quad v \in H_0^t(\Omega_i)$
>   with tolerance $\|e_i^{(n)} - \tilde{e}_i^{(n)}\| \leq \frac{1-\varrho}{2m\omega}\tilde{\varrho}^n M$.
>   **end for**
> $u^{(n+1)} := u^{(n)} + \omega \sum_{i=0}^{m-1} \tilde{e}_i^{(n)}$
> **end for**

This algorithm is convergent under the same assumption $\varrho < 1$, or, equivalently, $\tilde{\varrho} < 1$, as the exact additive Schwarz method. This is a consequence of the following result.

Proposition 4.1.8 ([82, Prop. 4.1]). *For the iterates $u^{(n)}$ from Algorithm 4, we have the error bound*

$$\|u^{(n)} - u\| \leq \tilde{\varrho}^n M.$$

Proof. The assertion is shown by induction, where the case $n = 0$ holds by definition of M and $u^{(0)}$. For the inductive step, assume the assertion holds for some $n \geq 0$. First of all, we separate the error made in the computation of $u^{(n+1)}$ from $u^{(n)}$ and write

$$\|u^{(n+1)} - u\| = \|u^{(n)} + \omega \sum_{i=0}^{m-1} \tilde{e}_i^{(n)} - u\|$$

$$\leq \|u^{(n)} + \omega \sum_{i=0}^{m-1} e_i^{(n)} - u\| + \omega\|\sum_{i=0}^{m-1}(\tilde{e}_i^{(n)} - e_i^{(n)})\|.$$

The first term corresponds exactly to one step of the *exact* additive Schwarz method starting from $u^{(n)}$. Hence, by the bound (4.18) and the tolerances for $\|e_i^{(n)} - \tilde{e}_i^{(n)}\|$, we obtain

$$\|u^{(n+1)} - u\| \leq \varrho\|u^{(n)} - u\| + \omega m \frac{1-\varrho}{2m\omega}\tilde{\varrho}^n M.$$

By the induction hypothesis, it holds that $\|u^{(n)} - u\| \leq \varrho^n M$. Hence, we have

$$\|u^{(n+1)} - u\| \leq \varrho\tilde{\varrho}^n M + \frac{1}{2}(1 - \varrho)\tilde{\varrho}^n M = \tilde{\varrho}^{n+1} M,$$

which completes the proof. $\qquad\square$

We have now laid out how the basic algorithms can be defined. However, we have not yet explained how the discretization with wavelets comes into play. This will be the task of the next sections, where we first introduce some essential tools we will need to handle numerically the discretized problem.

4.2 Building blocks

In this section, we introduce important tools needed to formulate an implementable adaptive wavelet version of the Schwarz algorithms. The task of these building blocks is to evaluate matrix-vector products, nonlinear expressions and the right-hand side in wavelet coordinates, at least up to given tolerances. Moreover, to make sure the algorithms are efficient in the sense that they keep a near-optimal balance between the degrees of freedom involved and the accuracy achieved, we will apply an adaptive thresholding strategy. These tools have already been developed over the course of time in various works, see, e.g., [6, 7, 25, 26, 27, 28, 75, 104]. For our purposes, it will be sufficient to recall their main ideas and properties. Since the thresholding step will also be involved in the construction of the other parts, we will start with this method.

4.2.1 Coarsening methods

The adaptive coarsening method we sketch now has the task to compute a sparse vector in aggregated tree structure that approximates a given input vector up to a given tolerance. This method has two main purposes. Firstly, working with sparse vectors will allow us to control the overall complexity of the algorithm and, secondly, tree structures are necessary when dealing with nonlinear functionals. This will be described later in more detail. Moreover, we want this method itself to manage this task with a number of operations that is roughly proportional to the support size of the input vector. This will be necessary to prove the optimality of our algorithm.

The first application of coarsening methods in adaptive wavelet methods were in algorithms for linear problems, where tree structured index sets do not necessarily have to be imposed. The basic idea from [25] for such a coarsening method is to simply sort the entries of the input vector by their absolute values and to remove as many of the smallest entries as the given tolerance allows. Since a complete sorting of the entries cannot be achieved in linear complexity, this idea has been refined in [6, 104]. The idea suggested there is to only sort the entries into buckets collecting entries with similar magnitudes. Then, entries from buckets representing the smallest magnitudes are removed from the vector. This way, the main properties of the coarsening method from [25] are retained while only requiring a number of operations proportional to the support size of the input vector.

Under the additional requirement that the output vector is supposed to be in aggregated tree structure, the construction of such an algorithm becomes substantially more involved. For the case of tree structures subject to wavelet bases, this problem was addressed in [27] with methods from [10]. For a more detailed description of the algorithmic aspects and further improvements, we also refer to [121]. The ideas then carry over to frames and aggregated trees in a straightforward manner, see [75]. To explain in detail this algorithm, let us first introduce some notation. Ideally, for $\mathbf{w} \in \ell_2(\Lambda_i)$, we would like to compute an ε-*best tree* $\mathcal{T}^* := \mathcal{T}^*(\mathbf{w}, \varepsilon)$, which is a tree such that

$$\|\mathbf{w} - \mathbf{w}_{|\mathcal{T}^*}\|_{\ell_2(\Lambda_i)} \leq \varepsilon$$

and $\#\mathcal{T}^*$ is minimal with this property. Finding such an ε-best tree in practice, however, is too expensive, because the number of possible trees grows exponentially with their size. Instead, the task is relaxed to computing an (ε, C)-*near-best tree*. For a parameter $C > 1$, this is a tree $\mathcal{T}$ such that $\|\mathbf{w} - \mathbf{w}_{|\mathcal{T}}\|_{\ell_2(\Lambda_i)} \leq \varepsilon$ holds with

$$\#\mathcal{T} \leq C \cdot \#\mathcal{T}^*(\mathbf{w}, \frac{\varepsilon}{C}).$$

This task can be achieved in linear time with the procedure from [10, 27] we will sketch now, see also [75, 121]. A key tool for this method is the error functional $h : \Lambda_i \to \mathbb{R}$, which for $\mathbf{v} \in \ell_2(\Lambda_i)$ and $\lambda \in \Lambda_i$ can be defined as

$$h(\lambda) := \tilde{v}_\lambda = \left(\sum_{\mu \succeq \lambda} |v_\mu|^2 \right)^{1/2}$$

and represents the part of the ℓ_2-norm of $\mathbf{v}$ that is caused by λ and all its successors. The basic idea is to start from a tree containing only the root nodes and to use this functional as an indicator where to refine the tree, which means that we add the children of indices to the tree, where the functional is largest. In this sense, the function h is also considered as an *energy functional* that becomes smaller or stays equal when the tree is refined. However, this approach taken by itself does not guarantee that the approximation error is indeed reduced by each refinement. That is because v_μ might vanish for all children μ of λ even though $h(\lambda)$ is large. Hence, it is suggested to apply a modified error functional $\tilde{h}(\lambda)$ that in fact decreases with each refinement. This functional is defined recursively. For the root nodes, we set $\tilde{h}(\lambda) := h(\lambda)$. If $\tilde{h}$ is defined for an index $\lambda \in \Lambda_i$, for each of its children $\mu \in \mathcal{C}(\lambda)$, we set

$$\tilde{h}(\mu) := \frac{\sum_{\eta \in \mathcal{C}(\lambda)} h(\eta)}{h(\lambda) + \tilde{h}(\lambda)} \tilde{h}(\lambda).$$

For a detailed analysis of this error functional, we refer to [10]. Note, however, that all siblings of an index are assigned the same value, so it has to be evaluated on only one of them. This modified functional $\tilde{h}$ is now used as the indicator where to refine the tree, whereas the original functional h serves as an error control. This results in the following method.

Algorithm 5 T_COARSE$[\mathbf{v}, \varepsilon] \mapsto \mathbf{w}_\varepsilon$

Compute $\mathcal{T}(\operatorname{supp} \mathbf{v})$, the smallest tree containing $\operatorname{supp} \mathbf{v}$.
Compute $h(\lambda)$, $\lambda \in \mathcal{T}(\operatorname{supp} \mathbf{v})$.
Let $\mathcal{T}$ be the set of all roots.
while $(\sum_{\lambda \in \mathcal{D}(\mathcal{T})} h(\lambda) > \varepsilon^2)$ **do**
    Add those $\lambda \in \mathcal{D}(\mathcal{T})$ to $\mathcal{T}$, where $\tilde{h}(\lambda)$ is largest.
end while
$\mathbf{w}_\varepsilon := \mathbf{v}_{|\mathcal{T}}$

Remark 4.2.1. *The smallest tree $\mathcal{T}(\operatorname{supp} \mathbf{v})$ containing $\mathbf{v} \in \ell_2(\Lambda_i)$ can be computed with $\mathcal{O}(\mathcal{T}(\# \operatorname{supp} \mathbf{v}))$ operations by a top-down-procedure, see [75]. This procedure can be sketched as follows. Starting from the maximum level*

$$j_{\max}(\mathbf{v}) := \max\{|\lambda|,\ \lambda \in \operatorname{supp} \mathbf{v}\}$$

we add to the index set all the parents of λ, $|\lambda| = j_{\max}(\mathbf{v})$. This procedure is repeated from $j_{\max}(\mathbf{v}) - 1$ all the way down to $j_{\min}(\mathbf{v}) + 1$, with $j_{\min}(\mathbf{v})$ defined analogously to $j_{\max}(\mathbf{v})$.

Remark 4.2.2. *As usual in the literature, we have formulated the coarsening method directly on the set Λ_i of wavelet indices. The algorithm can be slightly modified so that it works on the set of generator indices. Then, the output vector has the structure of a complete tree, see [76, Section 3.3] for details. As outlined before, completeness of a tree makes data structures easier to handle. Alternatively, one can also recomplete the tree in a post-processing step.*

The coarsening method can be generalized to aggregated trees $\mathcal{T} = \bigcup_{i=0}^{m-1} \{i\} \times \mathcal{T}_i$ by simply calling the coarsening method for each of the local trees with a modified tolerance. This leads to the following method, see [75, Remark 4.1].

Algorithm 6 AT_COARSE$[\mathbf{v}, \varepsilon] \mapsto \mathbf{w}_\varepsilon$

% Write $\mathbf{v} = (\mathbf{v}_0, \ldots, \mathbf{v}_{m-1})$ with $\mathbf{v}_i \in \{i\} \times \Lambda_i$.
for $i = 0, \ldots, m-1$ **do**
    $\mathbf{w}_i := $ T_COARSE$[\mathbf{v}_i, \frac{\varepsilon}{\sqrt{m}}]$
end for
$\mathbf{w}_\varepsilon := (\mathbf{w}_0, \ldots, \mathbf{w}_{m-1})$

The main properties of this method are summarized in the following *coarsening lemma*, which was shown for trees in [27] and generalized to aggregated trees in [75]. The lemma states that, if the input vector is close to another vector that lies in a given approximation class, the output vector from the coarsening method is contained in the same class, given that the tolerance is sufficiently large.

Lemma 4.2.3. *Let $\varepsilon > 0$, $\mathbf{w} \in \ell_2(\Lambda)$ with $\#\operatorname{supp}\mathbf{w} < \infty$ and $\mathbf{v} \in \mathcal{A}^s_{\mathcal{AT}}$. There exists a constant $C^* > 1$ such that under the condition $\|\mathbf{v} - \mathbf{w}\|_{\ell_2(\Lambda)} \leq \frac{\varepsilon}{2C^*+1}$, for the output vector*

$$\mathbf{w}_\varepsilon := \mathbf{AT\_COARSE}[\mathbf{w}, \frac{2C^*}{2C^*+1}\varepsilon],$$

we have

$$\|\mathbf{v} - \mathbf{w}_\varepsilon\|_{\ell_2(\Lambda)} \leq \varepsilon,$$
$$\#\operatorname{supp}\mathbf{w}_\varepsilon \lesssim \varepsilon^{-1/s}\|\mathbf{v}\|^{1/s}_{\mathcal{A}^s_{\mathcal{AT}}},$$
$$\|\mathbf{w}_\varepsilon\|_{\mathcal{A}^s_{\mathcal{AT}}} \lesssim \|\mathbf{v}\|_{\mathcal{A}^s_{\mathcal{AT}}},$$

and the number of operations needed to compute $\mathbf{w}_\varepsilon$ is proportional to $\#\mathcal{T}(\operatorname{supp}\mathbf{v})$.

Proof. See [75, Lemma 4.2] and [27, Propositions 6.2, 6.3]. □

Remark 4.2.4. *In some variants of Lemma 4.2.3 in the literature, another additional constant appears in the support estimates. In the above standard version, the additional terms are hidden in the estimates. Since we are interested in the case $\varepsilon \to 0$, thus $\varepsilon^{-1/s}\|\mathbf{v}\|^{1/s}_{\mathcal{A}^s_{\mathcal{AT}}} \to \infty$, additional constants are outweighted, so the distinction is of minor importance.*

In the numerical algorithms presented in Section 4.3, the coarsening method will serve two purposes. On the one hand, it will be applied after a number of iterations to reduce the support sizes of the iterates. On the other hand, it can be used combined with the following methods for matrix-vector products or the evaluation of the right-hand side to assure that their output is in tree structure.

4.2.2 Evaluation of matrix-vector products

In Section 2.3, we have seen that the application of the linear part $\mathbf{A} : \ell_2(\Lambda) \to \ell_2(\Lambda)$ can be written as an *infinite-dimensional* matrix-vector product $\mathbf{v} \mapsto \mathbf{Av}$. The evaluation of this mapping, at least up to a given error, is only possible in finite time if we have further information about the structure of the matrix. If the entries of the matrix decay sufficiently fast, there is hope that cut-off strategies can be applied so that we can work with a finite segment of the matrix without losing too much accuracy. Fortunately, the properties of many wavelet constructions, in particular their cancellation properties and locality, give rise to exactly these required decay estimates. The cut-off strategies and the resulting algorithms have been developed and analyzed in a series of works, see, e.g., [6, 25, 104]. The central assumption is that the matrix can be well approximated by sparse matrices. This property will be called *compressibility*.

Definition 4.2.5. *A matrix* $\mathbf{A}$ *that is bounded on* $\ell_2(\Lambda)$ *is called* s^*-*compressible if there exists a summable sequence* $(a_j)_{j\in\mathbb{N}}$ *and matrices* $\mathbf{A}_j$, $j \in \mathbb{N}$, *created by dropping entries from* $\mathbf{A}$, *with at most* $2^j a_j$ *entries in each row and column such that*

$$\|\mathbf{A} - \mathbf{A}_j\| \leq C_j$$

and $(2^{sj}C_j)_{j\in\mathbb{N}}$ *is summable for all* $s < s^*$. *Here and from now on, the matrix norm* $\|\cdot\|$ *is to be read as the operator norm on* $\ell_2(\Lambda)$.

Below, we will give a result that ensures the s^*-compressibility of matrices corresponding to a broad range of operators. If this property holds, the basic idea from [25] for the computation of the inexact matrix-vector product is to sort the entries of the input vector by their size and to match the largest entries of the input vector with matrices $\mathbf{A}_j$ with larger j. Since the matrices $\mathbf{A}_j$ become closer to $\mathbf{A}$ and more dense with increasing j, this means that a larger part of the computational effort is spent on the largest entries of the input vector. This approach balances well the costs for computing the approximation with the error. However, it involves a complete sorting of the input vector $\mathbf{v}$. This requires $\mathcal{O}(\# \operatorname{supp} \mathbf{v} \cdot \log(\# \operatorname{supp} \mathbf{v}))$ operations. Hence, to retain linear complexity in the degrees of freedom, we shall use a modified version of the algorithm introduced in [104]. By applying a binning strategy instead of the complete sorts, similarly to the coarsening method, this variant avoids the log-factors in the complexity estimates.

Algorithm 7 APPLY$[\mathbf{A}, \mathbf{v}, \varepsilon] \mapsto \mathbf{w}_\varepsilon$

- Let $q := \lceil \log(2/\varepsilon \cdot (\#\operatorname{supp}\mathbf{v})^{1/2}\|\mathbf{v}\|_{\ell_2(\Lambda)}\|\mathbf{A}\|) \rceil$.

- Partition the elements of $\mathbf{v}$ into bins V_i, where

$$V_i := \{v_\lambda, |v_\lambda| \in (2^{-i-1}\|\mathbf{v}\|_{\ell_2(\Lambda)}, 2^{-i}\|\mathbf{v}\|_{\ell_2(\Lambda)}]\}, \quad 0 \le i < q$$

  and the remaining elements are put into V_q.

- For $k = 0, 1, \ldots$, successively construct $\mathbf{v}_{[k]}$ by extracting $2^k - \lfloor 2^{k-1} \rfloor$ elements from V_0. If V_0 is empty, proceed with V_1 and so on. Stop if $\bigcup_i V_i$ becomes empty or

$$\|\mathbf{A}\|\,\|\mathbf{v} - \sum_{k=0}^{l} \mathbf{v}_{[k]}\|_{\ell_2(\Lambda)} \le \varepsilon/2,$$

  and set $l := k$.

- Choose $j \ge l$ minimal such that

$$\sum_{k=0}^{l} C_{j-k}\|\mathbf{v}_{[k]}\|_{\ell_2(\Lambda)} \le \varepsilon/2.$$

- For $0 \le k \le l$, compute the non-zero entries of $\mathbf{A}_{j-k}$ with a column index that appears in $\mathbf{v}_{[k]}$ and set

$$\mathbf{w}_\varepsilon := \sum_{k=0}^{l} \mathbf{A}_{j-k}\mathbf{v}_{[k]}.$$

The following result shows that this method indeed approximates the matrix-vector product up to any given tolerance. Moreover, if the matrix is s^*-compressible with a sufficiently high parameter s^*, the method preserves the approximation class of the input vector. Furthermore, if we tacitly assume that the entries of the matrix can be computed at unit cost, we can also control the overall complexity of the procedure.

Proposition 4.2.6 ([104, Prop. 3.8]). *Let $\mathbf{A}$ be s^*-compressible. Then, for any $\mathbf{v} \in \ell_2(\Lambda)$, $\varepsilon > 0$, the output $\mathbf{w}_\varepsilon := \mathbf{APPLY}[\mathbf{A}, \mathbf{v}, \varepsilon]$ fulfills*

$$\|\mathbf{Av} - \mathbf{w}_\varepsilon\|_{\ell_2(\Lambda)} \le \varepsilon.$$

Moreover, if $\mathbf{v} \in \ell_\tau^w(\Lambda)$, $\frac{1}{\tau} = s + \frac{1}{2}$, $s \in (0, s^*)$, we have

$$\#\operatorname{supp}\mathbf{w}_\varepsilon \lesssim \varepsilon^{-1/s}|\mathbf{v}|_{\ell_\tau^w(\Lambda)}^{1/s} \tag{4.19}$$

and the number of operations needed to compute $\mathbf{w}_\varepsilon$ is bounded by a constant times $\varepsilon^{-1/s}|\mathbf{v}|_{\ell_\tau^w(\Lambda)}^{1/s} + \#\operatorname{supp}\mathbf{v}$.

The output of the method **APPLY**, however, is not necessarily in aggregated tree structure. To achieve this, following [75], we combine it with the coarsening method from Subsection 4.2.1 and set

$$\mathbf{AT\_APPLY}[\mathbf{A}, \mathbf{v}, \varepsilon] := \mathbf{AT\_COARSE}[\mathbf{APPLY}[\mathbf{A}, \mathbf{v}, \frac{1}{2C^* + 1}\varepsilon], \frac{2C^*}{2C^* + 1}\varepsilon],$$

where C^* is the constant from Lemma 4.2.3. Let us summarize the properties of this method.

Proposition 4.2.7. *Assume that* $\mathbf{A}$ *is* s^*-*compressible and bounded on* $\mathcal{A}^s_{\mathcal{AT}}$, *where* $s \in (0, s^*)$. *Then, for* $\mathbf{v} \in \mathcal{A}^s_{\mathcal{AT}}$ *and* $\mathbf{w}_\varepsilon := \mathbf{AT\_APPLY}[\mathbf{A}, \mathbf{v}, \varepsilon]$, *it holds that*

$$\|\mathbf{Av} - \mathbf{w}_\varepsilon\|_{\ell_2(\Lambda)} \leq \varepsilon \tag{4.20}$$

$$\# \operatorname{supp} \mathbf{w}_\varepsilon \lesssim \varepsilon^{-1/s} \|\mathbf{v}\|_{\mathcal{A}^s_{\mathcal{AT}}}^{1/s} \tag{4.21}$$

$$\|\mathbf{w}_\varepsilon\|_{\mathcal{A}^s_{\mathcal{AT}}} \lesssim \|\mathbf{v}\|_{\mathcal{A}^s_{\mathcal{AT}}}. \tag{4.22}$$

Moreover, the number of operations needed to compute $\mathbf{w}_\varepsilon$ *is bounded by a constant multiple of* $\varepsilon^{-1/s} \|\mathbf{v}\|_{\mathcal{A}^s_{\mathcal{AT}}}^{1/s} + \#\mathcal{T}(\operatorname{supp} \mathbf{v})$.

Proof. The proof is essentially already given in [27, 75, 76]. The error bound (4.20) follows directly from Lemma 4.2.3 and Proposition 4.2.6. Moreover, the combination of these two results yields $\# \operatorname{supp} \mathbf{w}_\varepsilon \lesssim \varepsilon^{-1/s} \|\mathbf{Av}\|_{\mathcal{A}^s_{\mathcal{AT}}}^{1/s}$ and $\|\mathbf{w}_\varepsilon\|_{\mathcal{A}^s_{\mathcal{AT}}} \lesssim \|\mathbf{Av}\|_{\mathcal{A}^s_{\mathcal{AT}}}$. Using the boundedness assumption on $\mathbf{A}$ on $\mathcal{A}^s_{\mathcal{AT}}$, the estimates (4.21) and (4.22) follow. For the proof of the bound for the computational costs, we refer to [76, Prop. 4.15]. □

In Proposition 4.2.7, it is assumed that the matrix is s^*-compressible and bounded on $\mathcal{A}^s_{\mathcal{AT}}$. Fortunately, for most cases of practical interest, these assumptions can be verified, albeit for a possibly limited range of parameters. The technique to show compressibilty results is to establish off-diagonal decay estimates for the entries of the matrix, see, e.g., [25, 42]. Let us summarize some important compressibility and boundedness results in the following proposition.

Proposition 4.2.8. *Assume that there exists a* $\rho > 0$ *such that* $\mathcal{L}$ *is bounded from* $H^{t+t'}(\Omega)$ *to* $H^{-t+t'}(\Omega)$ *for all* $|t'| \leq \rho$. *Moreover, assume that the underlying wavelet basis allows a characterization* (2.14) *of Sobolev spaces for* $s \in (-\tilde{\gamma}, \gamma)$. *Then,* $\mathbf{A}$ *is* s^*-*compressible and bounded on* $\mathcal{A}^s_{\mathcal{AT}}$ *for all* $0 < s < s^* := \frac{2\sigma - d}{2d}$, *whenever* $\sigma \leq \rho$, $t + \sigma < \gamma$, $t - \sigma > -\tilde{\gamma}$, $\sigma \leq \frac{d}{2} + 2\tilde{l} + 2t$.

Proof. See [76, Remark 4.12] and [25, Prop. 3.4] for the compressibility and [76, Lemma 4.16] for the boundedness on $\mathcal{A}^s_{\mathcal{AT}}$. □

The central assumption there is that the operator $\mathcal{L}$ maps $H^{t+t'}(\Omega)$ boundedly into $H^{-t+t'}(\Omega)$ for $t' > 0$ in a nonempty range. For the standard examples we work with such as the Laplace operator, this assumption is fulfilled. However, the maximal s^* for which s^*-compressibility and boundedness on $\mathcal{A}^s_{\mathcal{AT}}$, $s \in (0, s^*)$ could be shown with this technique might stay below $\frac{l-t}{d}$, see also the discussion in [105]. Recalling the results presented in Chapter 3 and in view of Proposition 4.2.7, it would be preferable to show these properties up to $\frac{l-t}{d}$ so that the routine **AT_APPLY** does not hamper the overall rate of the algorithm. For suitable spline wavelet bases with sufficiently many vanishing moments, the results from [105] show that at least the higher compressibility can be guaranteed. These estimates can be generalized from wavelet bases to aggregated wavelet frames constructed from local spline wavelets of order l with $\tilde{l}$ vanishing moments, that are globally $(l-2)$-times continuously differentiable. Under a few additional assumptions, that are all fulfilled for standard spline wavelet constructions, see [110], we have the following compressibility result. This is a direct consequence of [110, Theorem 2.1] and the remarks thereafter.

Proposition 4.2.9. *Assume that the coefficients $a_{\alpha,\beta}$ in the differential operator (1.9) are sufficiently smooth. If $\tilde{l} \geq l - 2t$ and $\frac{l-t}{d} \geq \frac{1}{2}$, then the matrix $\mathbf{A}$ is s^*-compressible for an $s^* \geq \frac{l-t}{d}$. If even $\tilde{l} > l - 2t$ and $\frac{l-t}{d} > \frac{1}{2}$, then the matrix is s^*-compressible for an $s^* > \frac{l-t}{d}$.*

Remark 4.2.10. *Let us shortly remark that another problem to take into account is how the matrices $\mathbf{A}_j$ can be computed in practice at an average cost of one unit per entry. This property is called* computability. *We will not discuss this issue in detail, but remark that using properly chosen quadrature rules, this task was resolved for wavelet bases in [64]. The computation of the entries becomes more complicated when working with aggregated wavelet frames, where wavelets from different subdomains might intersect on the overlap in an irregular pattern. However, with an appropriate splitting of the matrix, this problem was tackled in [110].*

4.2.3 Evaluation of nonlinear expressions

In this subsection, we describe the approximate evaluation of the discretized nonlinearity $\mathbf{G}(\mathbf{v}) = (\langle \mathcal{G}(v), \psi_\lambda \rangle)_{\lambda \in \Lambda}$, $v = \mathbf{v}^\top \Psi^{H^t_0(\Omega)}$. A naive approach for the calculation of one coefficient would be to evaluate both $\mathcal{G}(v)$ and ψ_λ on a sufficiently fine grid and then use numerical integration. However, without further measures, it is not clear which coefficients are necessary to achieve a given overall accuracy. Furthermore, it is far from obvious how to control the computational effort so that for the overall algorithm, roughly linear complexity can be achieved. Instead, following [6, 7, 27, 28, 75, 76], we have to divide the evaluation into a two-step approach. In the first step, the *prediction* method, a tree-structured index set $\mathcal{T}$ is constructed such that $\|\mathbf{G}(\mathbf{v}) - \mathbf{G}(\mathbf{v})_{|\mathcal{T}}\|_{\ell_2(\Lambda)}$ is smaller than any prescribed tolerance and that the size of $\mathcal{T}$ is near-optimal. In the second step, the *recover* method, quadrature is performed on this index set, making intense use of its tree structure. In particular for the prediction step, it is inevitable to require that the coefficients fulfill a decay condition. To formulate this condition, let as usual $\Psi^{L_2(\Omega)}$ be the aggregated Gelfand frame for $(H_0^t(\Omega), L_2(\Omega), H^{-t}(\Omega))$ from Proposition 2.2.16 so that $\Psi^{H_0^t(\Omega)} = \{\psi_\lambda\}_{\lambda \in \Lambda} = \mathbf{D}^{-t}\Psi^{L_2(\Omega)}$ is a frame in $H_0^t(\Omega)$ with the non-canonical dual $\tilde{\Psi}^{H^{-t}(\Omega)} = \{\tilde{\psi}_\lambda\}$ in $H^{-t}(\Omega)$. Then, we assume that there exists a $\gamma > \frac{d}{2}$ such that for all $v = \sum_{\lambda \in \Lambda} \langle v, \tilde{\psi}_\lambda \rangle \psi_\lambda = \mathbf{v}^\top \Psi^{H_0^t(\Omega)}$ with finitely supported $\mathbf{v} \in \ell_2(\Lambda)$ and all $|\lambda| > j_0$, we have

$$|\langle \mathcal{G}(v), \psi_\lambda \rangle| \leq C(\|\mathbf{v}\|_{\ell_2(\Lambda)}) \sup_{\mu : \, \text{supp}\, \psi_\mu \cap \text{supp}\, \psi_\lambda \neq \emptyset} |\langle v, \tilde{\psi}_\mu \rangle| 2^{-\gamma(|\lambda| - |\mu|)}, \tag{4.23}$$

where $C(x)$ is a positive, nondecreasing function. For suitable wavelet bases, estimates of this kind can be shown if the nonlinearity has at most polynomial growth, as stated in the following result.

Lemma 4.2.11 ([75, Lemma 4.4]). *Assume that the wavelets ψ_λ are contained in C^k, $k \in \mathbb{N}$ and have k vanishing moments. Then, the estimate (4.23) holds for $\gamma := r + t + \frac{d}{2}$ with*

$$r := \begin{cases} \lceil \min\{k, p, n^*\} \rceil & \text{if } t < \frac{d}{2} \text{ and } G \text{ fulfills (1.14) for } 0 \leq p < \frac{d+2t}{d-2t}, \\ \min\{k, n^*\} & \text{if } t \geq \frac{d}{2} \text{ and } G \text{ fulfills (1.14) for some } p \geq 0. \end{cases}$$

Following the lines from [75], we are now ready to outline the prediction step. As a first part of this, let us define the following *expansion process*. Starting from an aggregated tree, this method generates a larger aggregated tree containing the original tree by adding wavelets on lower levels if their support intersects the support of a wavelet of higher level that is already part of the tree.

Algorithm 8 EXPAND$[\mathcal{T}] \mapsto \tilde{\mathcal{T}}$

> **for all** $\lambda \in \mathcal{T}$ **do**
> $\Phi_0(\lambda) := \{\lambda\}$
>   **for** $k = 1, \ldots, |\lambda|$ **do**
> $\Phi_k(\lambda) := \{\mu \in \Lambda, \, |\mu| = |\lambda| - k, \, \exists \xi \in \Phi_{k-1}(\lambda) \, \text{supp}\, \psi_\xi \cap \text{supp}\, \psi_\mu \neq \emptyset\}$
>   **end for**
> $\Phi(\lambda) := \bigcup_{k=0}^{|\lambda|} \Phi_k(\lambda)$
> **end for**
> $\tilde{\mathcal{T}} := \bigcup_{\lambda \in \mathcal{T}} \Phi(\lambda)$

The above process makes sure that all indices belonging to wavelets ψ_λ that intersect the support of a wavelet ψ_μ, $\mu \in \mathcal{T}$ with $|\mu| > |\lambda|$ are included in the tree. This feature is sometimes referred to as *expansion property*. Moreover, this method increases the size of the tree by at most a constant factor, see [75, Lemma 4.3]. The expansion procedure serves as a preparation for the prediction of the relevant indices of $\mathbf{G}(\mathbf{v})$, that, in turn, can be sketched as follows. Using coarsening with different tolerances, we filter out the most important coefficients in $\mathbf{v}$. These are usually the largest coefficients in $\mathbf{v}$, but we have to take into account the restriction to tree-structured index sets. Then, we add to these indices further indices on the next higher levels that correspond to wavelets with intersecting support. This reflects the off-diagonal decay assumed in the estimate (4.23), because indices on much higher levels are less vital for the approximation. Moreover, the more important an index in $\mathbf{v}$ is, the more neighbouring levels are taken into account. These principles are reflected in the following algorithm from [28, 75, 76]. Here, we understand the output of the coarsening method as an index set.

Algorithm 9 SUPPORT_PREDICTION$[\mathbf{v}, \varepsilon] \mapsto \check{\mathcal{T}}$

$\%$ *Let* $J := \min\{j \in \mathbb{N} : \frac{2^j \varepsilon}{j+1} \geq \|\mathbf{v}\|_{\ell_2(\Lambda)}\}$.
$\%$ *Let* $\alpha := \frac{2}{2\gamma - d}$ *with* γ *from* (4.23).
$\mathcal{T}_0 := \mathbf{EXPAND}[\mathbf{AT\_COARSE}[\mathbf{v}, \varepsilon]]$
for $j = J, \ldots, 1$ **do**
$\quad \mathcal{T}_j := \mathbf{EXPAND}[\mathbf{AT\_COARSE}[\mathbf{v}, \frac{2^j \varepsilon}{j+1}]]$
$\quad \Delta_{j-1} := \mathcal{T}_{j-1} \setminus \mathcal{T}_j$
$\quad$ **for all** $\mu \in \Delta_{j-1}$ **do**
$\quad\quad \Theta_{\varepsilon,\mu} := \{\lambda : \text{supp}(\psi_\lambda) \cap \text{supp}(\psi_\mu) \neq \emptyset, |\mu| \leq |\lambda| \leq |\mu| + (j-1)\alpha\}$
$\quad$ **end for**
end for
$\check{\mathcal{T}} := \mathcal{R} \cup \mathcal{T}_0 \cup \left(\bigcup_{\mu \in \mathcal{T}_0} \Theta_{\varepsilon,\mu} \right)$

The output of this method is an aggregated tree with a size that is asymptotically optimal. This fact is stated in the following result.

Theorem 4.2.12 ([75, Th. 4.1]). *The output $\check{\mathcal{T}}$ from Algorithm 9 is an aggregated tree with*

$$\|\mathbf{G}(\mathbf{v}) - \mathbf{G}(\mathbf{v})_{|\check{\mathcal{T}}}\|_{\ell_2(\Lambda)} \lesssim \varepsilon.$$

If $\mathbf{v} \in \mathcal{A}^s_{AT}$, $0 < s < \frac{2\gamma - d}{2d}$, we have

$$\#\check{\mathcal{T}} \lesssim \varepsilon^{-1/s} \|\mathbf{v}\|^{1/s}_{\mathcal{A}^s_{AT}} + 1.$$

Thus, it follows that $\mathbf{G}(\mathbf{v}) \in \mathcal{A}^s_{AT}$ with

$$\|\mathbf{G}(\mathbf{v})\|_{\mathcal{A}^s_{AT}} \lesssim \|\mathbf{v}\|_{\mathcal{A}^s_{AT}} + 1.$$

The number of operations needed to compute $\check{\mathcal{T}}$ is bounded by a constant multiple of $\varepsilon^{-1/s} \|\mathbf{v}\|^{1/s}_{\mathcal{A}^s_{AT}} + \#\mathcal{T}(\text{supp}\,\mathbf{v}) + 1$. The involved constants depend only on d, m, s and $\|\mathbf{v}\|_{\ell_2(\Lambda)}$.

Following [6, 75, 76], let us now turn to the second step in the evaluation of nonlinear expression. The task of this scheme called **RECOVER** is to calculate or approximate the coefficients $(\mathbf{G}(\mathbf{v}))_\lambda = \langle \mathcal{G}(v), \psi_\lambda \rangle$ on a given tree-structured index set $\lambda \in \mathcal{T}$. The algorithm is first explained for a pair of biorthogonal Riesz bases $\Psi^{L_2(\Omega)}$, $\tilde{\Psi}^{L_2(\Omega)}$ in $L_2(\Omega)$. The coefficients with respect to the rescaled basis in $H_0^t(\Omega)$ can then be obtained by scaling. Moreover, the process can be generalized to aggregated wavelet frames by a patchwise application of the local evaluation process. Both of these generalizations are taken into account in the error analysis.

To formulate the algorithm, let us first fix some notation. As usual, we denote by Δ_j the index set belonging to all generators on level j, $j \geq j_0$ and we write ∇_j, $j \geq j_0$ for the index set belonging to all wavelets on level j. For any $\lambda \in \Delta_j$, we then set $c_\lambda := \langle \phi_\lambda, \mathcal{G}(v) \rangle$, and, similarly, for $\lambda \in \bigcup_j \nabla_j$, we set $d_\lambda := \langle \psi_\lambda, \mathcal{G}(v) \rangle$. We collect these coefficients and write $\mathbf{c}_j(S) := (c_\lambda)_{\lambda \in S \cap \Delta_j}$ for a subset $S \subset \bigcup_j \Delta_j$ and define $\mathbf{d}_j(S)$ analogously. Moreover, we use the abbreviations $\mathbf{c}_j := \mathbf{c}_j(\Delta_j)$ and $\mathbf{d}_j := \mathbf{d}_j(\nabla_j)$. Furthermore, for a given tree $\mathcal{T}$, we denote by $t_j := \mathcal{T} \cap \nabla_j$ its indices corresponding to wavelets on level j.

At the core of the **RECOVER** scheme are the refinement matrices and the two-scale relation (2.10) from Lemma 2.2.4. This relation allows us to gradually trace back the calculation of the coefficients $\mathbf{d}_j(\mathcal{T})$ to the computation of some coefficients $\mathbf{c}_j$ involving generators. This idea is outlined in the following lemma. The results are verified in the proof of Proposition 1 and in Section 3.2 in [7], or in [76].

Lemma 4.2.13. *Let $\mathcal{T}$ be a tree, t_j° the generator indices corresponding to t_j, compare Subsection 3.4.1, and $\tilde{\mathbf{M}}_{j,k}^{(\lambda)}$ the λ-th column of the matrix $\tilde{\mathbf{M}}_{j,k}$, $k = 0, 1$. Define the index sets*

$$H_{j+1} := \bigcup_{\lambda \in t_j} \operatorname{supp} \tilde{\mathbf{M}}_{j,1}^{(\lambda)} \cup \bigcup_{\lambda \in t_j^\circ} \operatorname{supp} \tilde{\mathbf{M}}_{j,0}^{(\lambda)},$$

and the vector

$$\tilde{\mathbf{c}}_{j+1} := \mathbf{c}(H_{j+1}) - (\tilde{\mathbf{M}}_{j,0} \mathbf{c}(\Delta_j \setminus t_j^\circ))(H_{j+1}).$$

Then, we have the relation

$$\tilde{\Phi}_{j+1}^\top \tilde{\mathbf{c}}_{j+1} = \tilde{\Phi}_j^\top \mathbf{c}(t_j^\circ) + \tilde{\Psi}_j^\top \mathbf{d}_j(t_j).$$

Moreover, applying $\mathbf{M}_{j,0}^\top$ to $\tilde{\mathbf{c}}_{j+1}$ yields $\mathbf{c}_j(t_j^\circ)$ and applying $\mathbf{M}_{j,1}^\top$ to $\tilde{\mathbf{c}}_{j+1}$ gives $\mathbf{d}_j(t_j)$.

We are now ready to define the algorithm **RECOVER** in the form from [75]. The basic idea of this scheme is to apply the results from Lemma 4.2.13 to compute the coefficients $\mathbf{d}_j$ in the tree with a top-down-scheme. This means that, starting from the highest level in the tree, the coefficients can be computed recursively and only inner products with generators are required to get to the next lower level. Furthermore, an observation from [7] is used that it is sufficient to consider the index set $H_{j+1} \setminus t_j$ instead of H_{j+1} in the calculation of $\tilde{\mathbf{c}}_{j+1}$. This means that only generators *outside* the tree are required. Because the algorithm is directly formulated for aggregated wavelet frames, the vectors $\mathbf{c}_j$, $\mathbf{d}_j$, the sets t_j, H_{j+1} and the refinement matrices $\mathbf{M}_{j,k}$, $\tilde{\mathbf{M}}_{j,k}$

are equipped with another index representing the subdomain, e.g., $\mathbf{M}_{i,j,k}$ denotes the refinement matrix $\mathbf{M}_{j,k}$ for the wavelet basis on the i-th subdomain, $0 \leq i \leq m-1$.

Algorithm 10 RECOVER$[G(\cdot), \mathbf{v}, \mathcal{T}] \mapsto \mathbf{w}$

---

for $i = 0, \ldots, m-1$ do

    % Choose $j_i^* \in \mathbb{N}$ minimal such that $t_{i,j} = \emptyset$ for all $j \geq j_i^*$.

    % Calculate H_{i,j_i^*+1} and $\mathbf{q}_{i,j_i^*+1}(H_{i,j_i^*+1})$ as an approximation to $\mathbf{c}_{i,j_i^*+1}(H_{i,j_i^*+1})$

    % by quadrature.

    $\hat{\mathbf{c}}_{i,j_i^*+1} := \mathbf{0}$

    for $j = j_i^*, j_i^* - 1, \ldots, j_0$ do

        % Calculate $H_{i,j} \setminus t_{i,j}^\circ$ and $\mathbf{q}_{i,j}(H_{i,j} \setminus t_{i,j}^\circ)$ as approximation to $\mathbf{c}_{i,j}(H_{i,j} \setminus t_{i,j}^\circ)$.

        $\bar{\mathbf{c}}_{i,j+1} := \hat{\mathbf{c}}_{i,j+1} + \mathbf{q}_{i,j+1}(H_{i,j+1} \setminus t_{i,j+1}^\circ) - (\tilde{\mathbf{M}}_{i,j,0}\mathbf{q}_{i,j}(H_{i,j} \setminus t_{i,j}^\circ))(H_{i,j+1} \setminus t_{i,j+1}^\circ)$

        $\hat{\mathbf{c}}_{i,j} := \mathbf{M}_{i,j,0}^\top \bar{\mathbf{c}}_{i,j+1}$

        $\mathbf{d}_{i,j}^R := \mathbf{M}_{i,j,1}^\top \bar{\mathbf{c}}_{i,j+1}$

    end for

    $\mathbf{d}_{i,j_0-1}^R := \hat{\mathbf{c}}_{i,j_0}$

    $\mathbf{d}_i^R := \bigcup_{j=j_0-1}^{j_i^*} \mathbf{d}_{i,j}^R$

end for

$\mathbf{d}^R := \bigcup_{i=0}^{m-1} \mathbf{d}_i^R$

$\mathbf{w} := \mathbf{D}^{-t}\mathbf{d}^R$.

The main properties of this method are summarized in the following lemma, see [75, Remark 5.1, Lemma 5.3] or [76, Proposition 4.36] for a proof.

Lemma 4.2.14. *Assume that the quadrature in Algorithm 10 can be performed, on average, at constant cost per coefficient and with a sufficiently small error. Let the aggregated tree $\mathcal{T} = \bigcup_{i=0}^{m-1} \mathcal{T}_i$ be well-graded, which means that for all $j \geq j_0$ and $0 \leq i \leq m-1$, it holds that*

$$\left(\bigcup_{\lambda: \lambda^\circ \in H_{i,j} \setminus t_{i,j}^\circ} \tilde{\mathbf{M}}_{i,j,0}^{(\lambda)} \right) \cap t_{i,j+1}^\circ = \emptyset. \tag{4.24}$$

Assume moreover that for all $\mu \in \mathcal{T}$ and $j < |\mu|$, we have

$$\operatorname{supp} \tilde{\psi}_\mu \cap \left(\bigcup_{\lambda^\circ \in H_{i,j} \setminus t_{i,j}^\circ} \operatorname{supp} \tilde{\phi}_{(i,\lambda^\circ)} \right) = \emptyset. \tag{4.25}$$

Then, it is $\|\mathbf{G}(\mathbf{v})_{|\mathcal{T}} - \mathbf{w}\|_{\ell_2(\Lambda)} \lesssim \|\mathbf{G}(\mathbf{v})_{|\mathcal{T}} - \mathbf{G}(\mathbf{v})\|_{\ell_2(\Lambda)}$ and the number of operations needed by the algorithm is bounded by a constant multiple of $\#\mathcal{T}$.

To sum up, the lemma states that the method preserves the error and complexity estimates of the support prediction step. Let us shortly discuss the assumptions made in the lemma and the involved quadrature. The calculation of $c_\lambda = \langle \phi_\lambda, \mathcal{G}(v) \rangle$ is in fact an inner product of a spline with the nonlinearity applied to v. As outlined in [6, 7], if $G(\cdot)$ is polynomial, this is an integral over piecewise polynomials, which can in principle be calculated exactly with an integration formula of sufficiently high, constant order. Moreover, in the case of spline wavelets constructed from B-splines, the generators are particularly cheap to evaluate. This can further be exploited by rewriting v as a linear combination of generators of higher order, so only generators are involved in the quadrature. For details, we refer to [6, Section 3.3]. Regarding the assumption of well-gradedness, note that by subdividing leaves, a given tree can be transformed into a well-graded tree with a size that is at most a constant multiple of the original size, see [6, 7]. Therefore, this condition does not endanger the overall complexity estimates. Assumption (4.25) says that the support of dual generators outside the tree does not intersect with the support of dual wavelets on higher levels with indices belonging to the tree. Although this assumption is not automatically fulfilled, it can be assured by another *re-expansion* step in the algorithm. This step does not affect asymptotic results. For details on this, we refer to [75, 76].

Following [76], we can now combine the support prediction step with the recover step to finally obtain an evaluation method for the nonlinear part. To make sure that the output remains in $\mathcal{A}^s_{\mathcal{AT}}$ if the input vector does, a coarsening step is added at the end. The algorithm is then defined as follows.

Algorithm 11 EVAL$[G(\cdot), \mathbf{v}, \varepsilon] \mapsto \mathbf{w}_\varepsilon$

% Choose the constant $0 < \tilde{c} < 1$ according to Lemma 4.2.14 so that the
% overall error from the prediction and recovery step stays below $\frac{\varepsilon}{2C^*+1}$.
$\mathcal{T} := \textbf{SUPPORT\_PREDICTION}[\mathbf{v}, \frac{\tilde{c}\varepsilon}{2C^*+1}]$
$\tilde{\mathbf{w}} := \textbf{RECOVER}[G(\cdot), \mathbf{v}, \mathcal{T}]$
$\mathbf{w}_\varepsilon := \textbf{AT\_COARSE}[\tilde{\mathbf{w}}, \frac{\varepsilon 2C^*}{2C^*+1}]$

The properties of this method can be summarized in the following result, see [75, Th. 6.1] or [76, Prop. 4.36].

Proposition 4.2.15. *Let the decay assumption* (4.23) *hold and the conditions from Lemma* (4.2.14) *be fulfilled. Then, for the output* $\mathbf{w}_\varepsilon$ *from* **EVAL**, *it holds that*

$$\|\mathbf{G}(\mathbf{v}) - \mathbf{w}_\varepsilon\|_{\ell_2(\Lambda)} \leq \varepsilon, \tag{4.26}$$

$$\# \operatorname{supp} \mathbf{w}_\varepsilon \lesssim \varepsilon^{-1/s}(\|\mathbf{v}\|_{\mathcal{A}^s_{\mathcal{AT}}}^{1/s} + 1), \tag{4.27}$$

$$\|\mathbf{w}_\varepsilon\|_{\mathcal{A}^s_{\mathcal{AT}}} \lesssim \|\mathbf{v}\|_{\mathcal{A}^s_{\mathcal{AT}}} + 1. \tag{4.28}$$

Moreover, the number of operations needed to compute $\mathbf{w}_\varepsilon$ *is bounded by a constant multiple of* $\varepsilon^{-1/s}(\|\mathbf{v}\|_{\mathcal{A}^s_{\mathcal{AT}}}^{1/s} + 1) + \#\mathcal{T}(\operatorname{supp} \mathbf{v})$.

Proof. The error bound (4.26) follows directly from the choice of the constant $\tilde{c}$ in Algorithm 11 and Lemma 4.2.3. Moreover, by the same lemma, we have

$$\#\operatorname{supp} \mathbf{w}_\varepsilon \lesssim \varepsilon^{-1/s}\|\mathbf{G}(\mathbf{v})\|_{\mathcal{A}^s_{\mathcal{AT}}}^{1/s},$$

from which the estimate 4.27 follows by Theorem 4.2.12. In a similar fashion, the third estimate directly follows from Lemma 4.2.3 and Theorem 4.2.12 as well. To establish a bound for the computational effort of the algorithm, note that the total number of operations needed for the support prediction and the recover part can be bounded by a constant multiple of

$$\varepsilon^{-1/s}\|\mathbf{v}\|_{\mathcal{A}^s_{\mathcal{AT}}}^{1/s} + \#\mathcal{T}(\operatorname{supp} \mathbf{v}) + 1$$

according to Theorem 4.2.12 and Lemma 4.2.14. Since the output of the recover step is already in tree structure, the number of operations needed for the coarsening step is at most a constant times

$$\#\breve{\mathcal{T}} \lesssim \varepsilon^{-1/s}\|\mathbf{v}\|_{\mathcal{A}^s_{\mathcal{AT}}}^{1/s} + 1.$$

Since $\#\mathcal{T}(\operatorname{supp} \mathbf{v}) \gtrsim 1$, adding up both of these estimates yields the bound for the overall complexity of Algorithm 11. $\qquad\square$

4.2.4 Evaluation of the right-hand side

Let us now turn to the last remaining building block required for the algorithm, the approximation of the discrete right-hand side $\mathbf{f} = (\langle f, \psi_\lambda \rangle)_{\lambda \in \Lambda}$. Since the realization of this task strongly depends on the form of f, we have to make the standard assumption that there exists a method

$$\mathbf{AT\_RHS}[\varepsilon] \mapsto \mathbf{f}_\varepsilon,$$

which for any $\varepsilon > 0$, $\mathbf{f} \in \mathcal{A}^s_{\mathcal{AT}}$, $s \in (0, s^*)$ computes an approximation to $\mathbf{f}$ in aggregated tree structure with the properties

$$\|\mathbf{f}_\varepsilon - \mathbf{f}\|_{\ell_2(\Lambda)} \le \varepsilon, \tag{4.29}$$

$$\#\operatorname{supp} \mathbf{f}_\varepsilon \lesssim \varepsilon^{-1/s}\|\mathbf{f}\|_{\mathcal{A}^s_{\mathcal{AT}}}^{1/s}, \tag{4.30}$$

$$\|\mathbf{f}_\varepsilon\|_{\mathcal{A}^s_{\mathcal{AT}}} \lesssim \|\mathbf{f}\|_{\mathcal{A}^s_{\mathcal{AT}}}. \tag{4.31}$$

Moreover, we assume that this method requires at most a constant multiple of $\varepsilon^{-1/s}\|\mathbf{f}\|_{\mathcal{A}^s_{\mathcal{AT}}}^{1/s}$ operations. Assuming that f is sufficiently smooth, a standard approach for the practical realization of such a method is to precompute the most relevant entries of $\mathbf{f}$ approximately by numerical integration. On this vector, we can then apply coarsening up to the given tolerance, see, e.g., [27].

4.3 Adaptive multiplicative wavelet Schwarz method

In the following, the adaptive wavelet algorithms will be constructed, starting with the multiplicative Schwarz method. In these algorithms, compared to the versions from Section 4.1, the continuous iterates $u^{(n)} \in H_0^t(\Omega)$ will replaced by discrete counterparts $\mathbf{u}^{(n)} \in \ell_2(\Lambda)$ in wavelet frame coordinates, $u^{(n)} = F^*\mathbf{u}^{(n)} = (\mathbf{u}^{(n)})^\top \Psi$, where here and in the remainder of this section, $\Psi := \Psi^{H_0^t(\Omega)}$ is a wavelet frame for $H_0^t(\Omega)$. Recall that, in general, the discrete representation $\mathbf{u}^{(n)}$ of $u^{(n)}$ is not unique. This is also reflected by the decomposition (2.3) of the sequence space

$$\ell_2(\Lambda) = \operatorname{ran} F \oplus \ker F^*.$$

In particular, in this decomposition $\mathbf{u}^{(n)}$ might have parts in the kernel of F^*, which do not affect $u^{(n)}$. Hence, these parts, also called *redundancies*, do not endanger the convergence of the continuous iterates measured in the Sobolev norm. However, carrying along such redundancies in the course of the computation might affect the computational effort and reduce the efficiency of the algorithm. Since the subproblems can only be solved inexactly, it is not clear in advance how to avoid this problem without taking additional measures. Hence, we will use a known technique that explicitly removes the parts in $\ker F^*$ from the current iterates.

4.3.1 Controlling redundancies by a projection step

To tackle the problem with the redundancies outlined above, we will apply a technique that was first introduced in [104]. To make use of this strategy, let us assume we have at hand a partition of unity according to Definition 2.2.13. Then, we can decompose a given $v = \mathbf{v}^\top \Psi \in H_0^t(\Omega)$ into its *local parts* $v_i := \sigma_i v \in H_0^t(\Omega_i)$. Now, let $\mathbf{w}_i$ be the uniquely determined expansion coefficients of v_i in the local Riesz bases $\Psi^{(i)} := \Psi^{H_0^t(\Omega_i)}$ for $H_0^t(\Omega_i)$ and set $\mathbf{w} := (\mathbf{w}_0, \ldots, \mathbf{w}_{m-1}) \in \ell_2(\Lambda)$. This vector is just another representation of v because of

$$v = \sum_{i=0}^{m-1} v_i = \sum_{i=0}^{m-1} (\mathbf{w}_i)^\top \Psi^{(i)} = \mathbf{w}^\top \Psi. \tag{4.32}$$

The key point, however, is that after this procedure, elements in the kernel of F^* are removed. This can simply be seen, because $F^*\mathbf{v} = 0$ implies $v_i = 0$, which in turn implies $\mathbf{w} = 0$. This strategy can also be interpreted as the application of a linear projector, which can be written as an infinite-dimensional matrix. This and further properties are summarized in the following lemma, which is a collection of results from [82, 104, 123]. In the formulation of this lemma, for an index $\lambda \in \Lambda = \bigcup_{i=0}^{m-1} \{i\} \times \Lambda_i$, we write $p(\lambda) := i$ whenever $\lambda \in \{i\} \times \Lambda_i$, which means that λ corresponds to a wavelet from the i-th subdomain.

Lemma 4.3.1. *Let* $\tilde{\Psi}^{(i)} = \Psi^{H^{-t}(\Omega_i)} = \{\tilde{\psi}^{(i)}_\lambda\}_{\lambda \in \{i\} \times \Lambda_i}$ *denote the dual Riesz basis on the i-th subdomain Ω_i. We define the matrix $\mathbf{P}$ by*

$$(\mathbf{P})_{\lambda,\mu} := \langle \tilde{\psi}^{(i)}_\lambda, \sigma_i \psi_\mu \rangle, \quad \lambda, \mu \in \Lambda, \quad i = p(\lambda). \tag{4.33}$$

Then, $\mathbf{P}$ is a bounded linear operator on $\ell_2(\Lambda)$. Moreover, it is a projection, $\mathbf{P}^2 = \mathbf{P}$, and $\ker \mathbf{P} = \ker F^*$. In addition, we have the norm equivalences

$$\|\mathbf{P}\mathbf{v}\|_{\ell_2(\Lambda)} \eqsim \|v\|_{H^t(\Omega)} \eqsim \|v\|, \quad \mathbf{v} \in \ell_2(\Lambda), \; v = \mathbf{v}^\top \Psi. \tag{4.34}$$

Proof. Let $\mathbf{v} \in \ell_2(\Lambda)$ and $v = \mathbf{v}^\top \Psi$. Then, we have

$$\mathbf{w}_i := (\mathbf{P}\mathbf{v})_{|\{i\} \times \Lambda_i} = \Big(\sum_{\mu \in \Lambda} \langle \tilde{\psi}^{(i)}_\lambda, \sigma_i \psi_\mu \rangle v_\mu \Big)_{\lambda \in \{i\} \times \Lambda_i} = \big(\langle \tilde{\psi}^{(i)}_\lambda, \sigma_i v \rangle \big)_{\lambda \in \{i\} \times \Lambda_i} \tag{4.35}$$

Hence, by Theorem 2.1.2, it holds that $\mathbf{w}_i^\top \Psi^{(i)} = \sigma_i v$. Using that the $\Psi^{(i)}$ are Riesz bases and the properties of the partition of unity, we have

$$\|\mathbf{w}_i\|_{\ell_2(\{i\} \times \Lambda_i)} \eqsim \|v_i\|_{H^t(\Omega_i)} \lesssim \|v\|_{H^t(\Omega)} \lesssim \|\mathbf{v}\|_{\ell_2(\Lambda)},$$

thus $\mathbf{P}$ is bounded. In addition, because of $\|\mathbf{P}\mathbf{v}\|_{\ell_2(\Lambda)} \eqsim \sum_{i=0}^{m-1} \|\mathbf{w}_i\|_{\ell_2(\Lambda_i)}$ and the norm equivalence between the energy norm and the Sobolev norm, the estimates (4.34) hold. We have already seen above from Equation (4.32) that $F^*\mathbf{P}\mathbf{v} = F^*\mathbf{v}$. Since by Equation (4.35), the image $\mathbf{P}\mathbf{v}$ only depends on $v = F^*\mathbf{v}$ and not on $\mathbf{v}$ itself, this shows that $\mathbf{P}^2 = \mathbf{P}$ and that $\ker F^* \subset \ker \mathbf{P}$. The converse inclusion holds as well, because $\mathbf{P}\mathbf{v} = 0$ implies $\sigma_i v = 0$ for all i, which gives $v = F^*\mathbf{v} = 0$. $\quad\square$

Let us now discuss the practical realization of the projector $\mathbf{P}$. The most natural idea is to approximate $\mathbf{P}\mathbf{v}$, $\mathbf{v} \in \ell_2(\Lambda)$ via an approximate matrix-vector multiplication like the method **APPLY** from Subsection 4.2.2. In [104, Section 4.5], it is shown that if the primal wavelets of order l fulfill a Bernstein estimate up to $\gamma > 0$, the matrix $\mathbf{P}$ is s^*-compressible with

$$s^* = \begin{cases} \min\{\frac{\gamma-t}{d-1}, \frac{l-t}{d}\} & , d > 1 \\ l - t & , d = 1 \end{cases}.$$

If $\mathbf{P}$ is furthermore bounded on $\mathcal{A}^s_{AT}$, we also have asymptotic optimality with respect to aggregated tree approximation, see Proposition 4.2.7. Below, we will discuss this assumption in a little more detail. For a practical application of the method **APPLY**, it is necessary to compute the relevant entries of $\mathbf{P}$. This is a severe difficulty because, other than usual, the dual Riesz bases from the subdomains are explicitly involved. These dual wavelets can be computed, but they often have a less convenient structure than the primal wavelets. For instance, for spline wavelets such as the construction from [95] used in our numerical experiments, the dual wavelets no longer have a spline structure. This significantly complicates the computation of the inner products in

Equation (4.33), that are needed for setting up the matrix. This is because, on the one hand, the evaluation of the dual wavelets is more expensive in practice and, on the other hand, the error made by the numerical integration is more difficult to control than when integrating over splines, where a composite integration rule of sufficiently high order even gives exact results. In [123, Section 6.2], a way to circumvent this problem was suggested, that we outline in the following.

To explain this strategy, recall that the application of the projector $\mathbf{P}$ amounts to computing the expansion coefficients of $v_i = \sigma_i v$ in terms of $\Psi^{(i)}$. Hence, the computation of $\mathbf{w}_i = (\mathbf{P}v)_{|\{i\}\times\Lambda_i}$ is equivalent to solving $(\Psi^{(i)})^\top \mathbf{w}_i = \sigma_i v$. Using that the $\Psi^{(i)}$ are Riesz bases for $H_0^t(\Omega_i)$, this is again equivalent to the linear system

$$\left(\sum_{\lambda \in \{i\}\times\Lambda_i} \langle \psi_\lambda, \psi_\mu \rangle w_\lambda \right)_{\mu \in \{i\}\times\Lambda_i} = (\langle \sigma_i v, \psi_\mu \rangle)_{\mu \in \{i\}\times\Lambda_i}. \tag{4.36}$$

In this system, the left-hand side can be read as a matrix-vector multiplication of the matrix $(\langle \psi_\lambda, \psi_\mu \rangle)_{\mu,\lambda \in \{i\}\times\Lambda_i}$ with the vector $\mathbf{w}_i$, where the matrix simply corresponds to the discretization of the identity operator. Hence, these simpler subproblems can be treated with a standard adaptive wavelet solver for linear problems. Moreover, since the identity operator is a simple operator of order zero, we expect these solvers to converge very fast. In addition, the right-hand side in Equation (4.36) can also simply be computed as a rectangular matrix-vector product with the input vector $\mathbf{v}$, because it is

$$(\langle \sigma_i v, \psi_\mu \rangle)_{\mu \in \{i\}\times\Lambda_i} = \left(\sum_{\lambda \in \Lambda} \langle \sigma_i \psi_\lambda, \psi_\mu \rangle v_\lambda \right)_{\mu \in \{i\}\times\Lambda_i}$$
$$= (\langle \sigma_i \psi_\lambda, \psi_\mu \rangle)_{\mu \in \{i\}\times\Lambda_i, \lambda \in \Lambda} \cdot \mathbf{v}.$$

This alternative approach for the application of the projector was further studied in [81]. There, it was verified that the above matrices involved in the subproblems are sufficiently compressible. Moreover, it was shown that this technique can be embedded into adaptive wavelet frame algorithms for linear problems such as the Richardson iteration from [104] and that optimal convergence rates are preserved. Although the general functionality of this idea could be confirmed by numerical experiments, it was observed that the effort for this application of $\mathbf{P}$ does not pay off in most cases. Hence, for the *practical* realization of adaptive wavelet frame algorithms, the projection step is usually omitted. Nevertheless, it may be needed for the theoretical verification of asymptotic optimality. Hence, we will incorporate the application of this step into the algorithms we are about to design. For simplicity, we will from now on use the formulation of the projection step as a matrix-vector product, despite the fact that by the above remarks, the practical realization should be implemented in a different way.

Remark 4.3.2. *So far, we have not yet been able to strictly verify that $\mathbf{P}$ is bounded on $\mathcal{A}^s_{\mathcal{AT}}$, although we expect this assumption to hold. One idea is to use Remark 3.3.3 to see that $\mathbf{v} \in \mathcal{A}^s_{\mathcal{AT}}$ implies a Besov regularity of v in the adaptivity scale. Then, we would like to use Proposition 3.4.7 to conclude that $\mathbf{P}v$, which are exactly the expansion coefficients of v with respect to the non-canonical dual frame, are again contained in $\mathcal{A}^s_{\mathcal{AT}}$. However, for this implication, regularity of v in a slightly stricter scale of Besov spaces is required. An alternative is to use the projector we will introduce in Chapter 6, which, however, requires stronger assumptions on the domain decomposition and the wavelet construction.*

4.3.2 The adaptive method and its convergence

We are now ready to define the adaptive multiplicative Schwarz method and prove its convergence, initially under the assumption that the local subproblems can be solved up to any given accuracy. The numerical treatment of the subproblems, and thus the verification of this assumption, will be addressed thereafter. To formulate the algorithm, we have to fix some additional constants. Because Ψ is a frame, the operator F^* is bounded. Together with the fact that the energy norm is equivalent to the Sobolev norm on $H^t_0(\Omega)$, this implies the existence of a constant $K_1 > 0$ such that

$$\|\mathbf{v}^\top \Psi\| \le K_1 \|\mathbf{v}\|_{\ell_2(\Lambda)} \tag{4.37}$$

holds for all $\mathbf{v} \in \ell_2(\Lambda)$. Furthermore, by Lemma 4.3.1, there exists a constant $K_2 > 0$, so that for all $\mathbf{v} \in \ell_2(\Lambda)$, we have the estimate

$$\|\mathbf{P}\mathbf{v}\|_{\ell_2(\Lambda)} \le K_2 \|\mathbf{v}^\top \Psi\|. \tag{4.38}$$

In the algorithm defined below, boldface characters as usual represent discrete iterates. The local subproblems are formulated in the continuous form, but they will later be solved in discretized form with respect to the local wavelet bases. Hence, it is safe to assume that the expansion coefficients of the output of the local solvers are always available.

Algorithm 12 MultSchw3 $[\varepsilon]$

% Let $M \geq \|u\|$.

% Let C^* be the constant from Lemma 4.2.3.

% Let $l^* \in \mathbb{N}$ be minimal such that $\tilde{\rho}^{l^*} \leq \frac{1}{2K_1K_2}\frac{1}{2C^*+1}\tilde{\rho}$.

% Let $\varepsilon_n := \tilde{\rho}^n M$, $n \in \mathbb{N}$.

$u^{(0)} := 0$

$n := 0$

while $\varepsilon_n > \varepsilon$ **do**

  $v_0^{(n)} := u^{(n)}$

  for $l = 0, \ldots, l^* - 1$ **do**

    $w_0^{(n,l)} := v^{(n,l)}$

    for $i = 0, \ldots, m - 1$ **do**

      Compute $\tilde{e}_i^{(n,l)} \in H_0^t(\Omega_i)$ as an approximation to $e_i^{(n,l)} \in H_0^t(\Omega_i)$ from

        $a(e_i^{(n,l)}, v) = -a(w_i^{(n,l)}, v) + f(v) - \mathcal{G}(v^{(n,l)})(v), \quad v \in H_0^t(\Omega_i)$

      with tolerance $\|\tilde{e}_i^{(n,l)} - e_i^{(n,l)}\| \leq \frac{1}{2m}\varepsilon_n\tilde{\rho}^l(1 - \rho)$.

      $w_{i+1}^{(n,l)} := w_i^{(n,l)} + \tilde{e}_i^{(n,l)}$

    end for

    $v^{(n,l+1)} := w_m^{(n,l)}$

  end for

  $\tilde{\mathbf{u}}^{(n+1)} := \mathbf{AT\_APPLY}[\mathbf{P}, \mathbf{v}^{(n,l^*)}, \frac{1}{2K_1}\frac{1}{2C^*+1}\varepsilon_{n+1}]$

  $\mathbf{u}^{(n+1)} := \mathbf{AT\_COARSE}[\tilde{\mathbf{u}}^{(n+1)}, \frac{1}{K_1}\frac{2C^*}{2C^*+1}\varepsilon_{n+1}]$

  $n := n + 1$

end while

$\mathbf{u}_\varepsilon := \mathbf{u}^{(n)}$

$u_\varepsilon := \mathbf{u}_\varepsilon^\top \Psi$

In a first step, we show that this method is convergent if the nonlinearity is sufficiently small.

Proposition 4.3.3. *Assume that $\rho < 1$. Then, for any given $\varepsilon > 0$, Algorithm 12 is convergent, i.e., it terminates after finitely many iterations with*

$$\|u^{(n)} - u\| \leq \varepsilon_n,$$

and, in particular,

$$\|u_\varepsilon - u\| \leq \varepsilon.$$

Proof. We show the first statement by induction, the second assertion then follows directly. The case $n = 0$ is clear by definition of M and $u^{(0)}$. Assume that the estimate holds for some $n \geq 0$. Then, analogously to the proof of Proposition 4.1.5, we see that one inner iteration reduces the error by a factor of $\tilde{\rho}$. Hence, using the induction hypothesis, we obtain

$$\|v^{(n,l)} - u\| \leq \tilde{\rho}^l\|u^{(n)} - u\| \leq \varepsilon_n\tilde{\rho}^l. \tag{4.39}$$

By the tolerance with which **AT_APPLY** is called and the choice of K_2 in (4.38), it holds that

$$\|\tilde{\mathbf{u}}^{(n+1)} - \mathbf{Pu}\|_{\ell_2(\Lambda)} \leq \|\mathbf{Pv}^{(n,l^*)} - \tilde{\mathbf{u}}^{(n+1)}\|_{\ell_2(\Lambda)} + \|\mathbf{Pv}^{(n,l^*)} - \mathbf{Pu}\|_{\ell_2(\Lambda)}$$
$$\leq \frac{1}{2K_1}\frac{1}{2C^*+1}\varepsilon_{n+1} + K_2\|v^{(n,l^*)} - u\|.$$

Combining this with the estimate (4.39) and the choice of l^* within Algorithm 12, we obtain

$$\|\tilde{\mathbf{u}}^{(n+1)} - \mathbf{Pu}\|_{\ell_2(\Lambda)} \leq \frac{1}{2K_1}\frac{1}{2C^*+1}\varepsilon_{n+1} + K_2\varepsilon_n\tilde{\rho}^{l^*} \leq \frac{1}{K_1}\frac{1}{2C^*+1}\varepsilon_{n+1}. \qquad (4.40)$$

Hence, by the choice of K_1 from (4.37) and with the tolerance for the coarsening step in mind, we end up with

$$\|u^{(n+1)} - u\| \leq K_1\|\mathbf{u}^{(n+1)} - \mathbf{Pu}\|_{\ell_2(\Lambda)}$$
$$\leq K_1(\|\tilde{\mathbf{u}}^{(n+1)} - \mathbf{u}^{(n+1)}\|_{\ell_2(\Lambda)} + \|\tilde{\mathbf{u}}^{(n+1)} - \mathbf{Pu}\|_{\ell_2(\Lambda)})$$
$$\leq K_1(\frac{1}{K_1}\frac{2C^*}{2C^*+1}\varepsilon_{n+1} + \frac{1}{K_1}\frac{1}{2C^*+1}\varepsilon_{n+1})$$
$$= \varepsilon_{n+1},$$

which shows the assertion. □

4.3.3 Solution of the local subproblems

Let us now turn to the problem of how to solve for $e_i^{(n,l)} \in H_0^t(\Omega_i)$ the linear subproblems

$$a(e_i^{(n,l)}, v) = -a(w_i^{(n,l)}, v) + f(v) - \mathcal{G}(v^{(n,l)})(v), \quad v \in H_0^t(\Omega_i) \qquad (4.41)$$

arising in the multiplicative Schwarz method. Recall that, already from the construction of the aggregated wavelet frame, we have available Riesz bases $\Psi^{(i)}$ for $H_0^t(\Omega_i)$. Therefore, the above equation is equivalent to the discretized linear subproblem

$$\mathbf{A}^{(i,i)}\mathbf{e}_i^{(n,l)} = -(\mathbf{Aw}_i^{(n,l)})_{|\Lambda_i} + \mathbf{f}_{|\Lambda_i} - \mathbf{G}(\mathbf{v}^{(n,l)})_{|\Lambda_i}. \qquad (4.42)$$

In particular, with some care being needed for the computation of the right-hand side, this problem could in principle be solved using standard wavelet methods for linear problems such as those presented in [25, 26]. However, since we would like to embed this local solver into the algorithm and prove overall complexity estimates, we have to be a little more specific. Our method of choice for the solution of the subproblems is a Richardson iteration from [26], which, in similar forms, has also been applied as a local solver in [82, 111, 123]. In practical applications, one could apply a Galerkin method following [25] instead, which might lead to a better quantitative performance, compare also the implementation in [111, 123]. Nevertheless, in this thesis, we restrict ourselves to the Richardson iteration. This is because the presentation and the analysis of this method is clearer and it involves fewer additional parameters that have to be estimated. Furthermore, we have observed that the by far largest share of the computational effort for the solution of the subproblems is spent on the approximation of the right-hand side in (4.42). Therefore, the speed-up from using a more complex local solver probably does not justify the significantly more technical presentation of an adaptive Galerkin method.

To specify the local solvers and to estimate their computational complexity, the following lemma will be of central importance. It allows us to estimate the energy norm of the solutions to the local subproblems. With the help of this result, we will later be able to show that in the local Richardson iterations, we only need a constant number of inner iterations to end up within the desired tolerance for the error.

Lemma 4.3.4. *For the exact solutions $e_i^{(n,l)}$ of the local subproblems in Algorithm 12, it holds that $\|e_i^{(n,l)}\| \lesssim \varepsilon_n \tilde{\rho}^l$ with a constant independent of n, i and l.*

Proof. Analogously to Equation (4.9), the solutions of the subproblems can be written as

$$e_i^{(n,l)} = P_i(u - w_i^{(n,l)}) + \mathcal{L}_i^{-1}(\mathcal{G}(v^{(n,l)}) - \mathcal{G}(u)).$$

Therefore, using that both the orthogonal projector P_i and $\mathcal{L}_i^{-1}$ have an operator norm of at most one in the energy norm and applying the error bound (4.39), we obtain

$$\|e_i^{(n,l)}\| \le \|u - w_i^{(n,l)}\| + c\|v^{(n,l)} - u\| \le \|u - w_i^{(n,l)}\| + c\varepsilon_n \tilde{\rho}^l.$$

We are now going to find an upper bound for the first term. Analogously to the estimate (4.15), we can calculate the error for the inner iterates by

$$w_i^{(n,l)} - u = (I - P_{i-1}) \cdot \ldots \cdot (I - P_0)(v^{(n,l)} - u)$$

$$- \sum_{j=0}^{i-1}(I - P_{i-1}) \cdot \ldots \cdot (I - P_{j+1})[\mathcal{L}_j^{-1}(\mathcal{G}(v^{(n,l)}) - \mathcal{G}(u)) + \delta_j^{(n,l)}].$$

with $\delta_j^{(n,l)} := \tilde{e}_j^{(n,l)} - e_j^{(n,l)}$. Recall that $\|(I - P_i)\| \le 1$, because $(I - P_i)$ is the orthogonal projector on the orthogonal complement of $H_0^t(\Omega_i)$ in $H_0^t(\Omega)$. Thus, for each $i \le m$, it is

$$\|w_i^{(n,l)} - u\| \le \|v^{(n,l)} - u\| + \sum_{j=0}^{i-1} c\|v^{(n,l)} - u\| + \|\delta_j^{(n,l)}\|$$

$$\le (1 + mc)\|v^{(n,l)} - u\| + \sum_{j=0}^{i-1} \frac{1}{2m}\varepsilon_n \tilde{\rho}^l(1 - \rho)$$

$$\le (1 + mc)\tilde{\rho}^l \varepsilon_n + m \cdot \frac{1}{2m}\varepsilon_n \tilde{\rho}^l(1 - \rho)$$

$$\le (1 + mc + \frac{1}{2}(1 - \rho))\varepsilon_n \tilde{\rho}^l.$$

Altogether, we finally obtain

$$\|e_i^{(n,l)}\| \le (1 + mc + \frac{1}{2}(1 - \rho) + c)\varepsilon_n\tilde{\rho}^l. \tag{4.43}$$

$\square$

Recall now, that in the adaptive multiplicative Schwarz algorithm, the local sub-problems are to be solved up to a tolerance $\frac{1}{2m}\varepsilon_n\tilde{\rho}^l(1 - \rho)$, where l is between 0 and $l^* - 1$. Hence, in view of the estimate (4.43), starting from the zero function as an initial guess for the solution of the local subproblems, it is sufficient to reduce the error by the constant factor R^{-1} in the energy norm, where

$$R := \frac{(1 + mc + \frac{1}{2}(1 - \rho) + c)\varepsilon_n\tilde{\rho}^l}{\frac{1}{2m}\varepsilon_n\tilde{\rho}^l(1 - \rho)} = \frac{2m(1 + mc + \frac{1}{2}(1 - \rho) + c)}{(1 - \rho)}.$$

Because the $\Psi^{(i)}$ are Riesz bases and in view of Lemma 2.3.3, the energy norm is equivalent to the $\|\cdot\|_{\mathbf{A}^{(i,i)}}$-norm of the expansion coefficients. Hence, it is also sufficient to reduce the error in this discrete norm by a constant factor, which we will denote by $\tilde{R}$.

Let us now sketch the principles of the Richardson iteration. In Lemma 2.3.3, we have seen that the main diagonal blocks $\mathbf{A}^{(i,i)}$ of the matrix $\mathbf{A}$ are positive definite. This allows us to make use of the following well-known result, compare, e.g., [100, 111], which lies at the core of the Richardson iteration.

Lemma 4.3.5. *For $0 < \alpha < \frac{2}{\lambda_{\max}(\mathbf{A}^{(i,i)})}$, it holds that*

$$\|\mathbf{I} - \alpha\mathbf{A}^{(i,i)}\|_{\mathbf{A}^{(i,i)}} < 1,$$

where the norm is to be read as the operator norm on $\ell_2(\Lambda_i)$ induced by $\|\cdot\|_{\mathbf{A}^{(i,i)}}$.

Proof. Because $\mathbf{A}^{(i,i)}$ is positive definite and symmetric with respect to both $\langle\cdot,\cdot\rangle_{\ell_2(\Lambda)}$ and $\langle\cdot,\cdot\rangle_{\mathbf{A}^{(i,i)}}$, it holds that

$$\|\mathbf{I} - \alpha\mathbf{A}^{(i,i)}\|_{\mathbf{A}^{(i,i)}} = \rho(\mathbf{I} - \alpha\mathbf{A}^{(i,i)}) = \max\{1 - \alpha\lambda_{\min}(\mathbf{A}^{(i,i)}), \alpha\lambda_{\max}(\mathbf{A}^{(i,i)}) - 1\},$$

which is smaller than one for all $0 < \alpha < \frac{2}{\lambda_{\max}(\mathbf{A}^{(i,i)})}$. $\square$

For a given right-hand side $\mathbf{r} \in \ell_2(\Lambda_i)$ and a starting vector $\mathbf{y}^{(0)}$, the exact iteration is now defined by

$$\mathbf{y}^{(r+1)} := \mathbf{y}^{(r)} + \alpha(\mathbf{r} - \mathbf{A}^{(i,i)}\mathbf{y}^{(r)}).$$

The error in the $(r + 1)$st step can be written as

$$\begin{aligned}
\mathbf{y}^{(r+1)} - (\mathbf{A}^{(i,i)})^{-1}\mathbf{r} &= \mathbf{y}^{(r)} + \alpha(\mathbf{r} - \mathbf{A}^{(i,i)}\mathbf{y}^{(r)}) - (\mathbf{A}^{(i,i)})^{-1}\mathbf{r} \\
&= (\mathbf{I} - \alpha\mathbf{A}^{(i,i)})(\mathbf{y}^{(r)} - (\mathbf{A}^{(i,i)})^{-1}\mathbf{r}). \tag{4.44}
\end{aligned}$$

Thus, in each step, we have at least a geometric reduction of the error measured in $\|\cdot\|_{\mathbf{A}^{(i,i)}}$ by the factor $\|\mathbf{I} - \alpha\mathbf{A}^{(i,i)}\|_{\mathbf{A}^{(i,i)}}$, which, by the above lemma, is smaller than one. In practice, these steps can only be carried out inexactly. With properly chosen tolerances, the geometric convergence can however be preserved. Adapting the local solvers from [82, 123], we propose the following algorithm for the solution of the local subproblems.

Algorithm 13 LocSolve[δ]

$\%$ *Choose* $\alpha, \xi > 0$ *so that* $\|\mathbf{I} - \alpha\mathbf{A}^{(i,i)}\|_{\mathbf{A}^{(i,i)}} \leq \xi < 1.$
$\%$ *Choose* $p \in \mathbb{N}$ *minimal with* $2\xi^p(\tilde{R} + \frac{1}{2}) \leq \frac{1}{2}.$
$tol := \frac{1}{6}\|(\mathbf{A}^{(i,i)})^{-1}\|_{\ell_2(\Lambda_i)\to\ell_2(\Lambda_i)}^{-1/2}\delta$
$tol_2 := \dfrac{\xi^p(\tilde{R}+\frac{1}{2})\delta}{\alpha p\|\mathbf{A}^{(i,i)}\|_{\ell_2(\Lambda_i)\to\ell_2(\Lambda_i)}^{1/2}}$
$\tilde{\mathbf{r}} := -\,\mathbf{AT\_APPLY}[\mathbf{A}, \mathbf{w}_i^{(n,l)}, tol]_{|\Lambda_i} + \mathbf{AT\_RHS}[\mathbf{f}, tol]_{|\Lambda_i} - \mathbf{EVAL}[\mathbf{G}, \mathbf{v}^{(n,l)}, tol]_{|\Lambda_i}$
$\mathbf{y}^{(0)} := \mathbf{0}$
for $r = 0, \dots, p-1$ **do**
$\quad \mathbf{y}^{(r+1)} := \mathbf{y}^{(r)} + \alpha(\tilde{\mathbf{r}} - \mathbf{AT\_APPLY}[\mathbf{A}^{(i,i)}, \mathbf{y}^{(r)}, tol_2])$
end for
$\tilde{\mathbf{e}}_i^{(n,l)} := \mathbf{y}^{(p)}$
$\tilde{e}_i^{(n,l)} := (\tilde{\mathbf{e}}_i^{(n,l)})^\top \Psi^{(i)}$

The convergence of this method is shown in the following lemma, which is a version of [82, Lemma 4.7] and [123, Lemma 6.3].

Lemma 4.3.6. *Let* $\delta > 0$ *and assume that for the exact solution* $e_i^{(n,l)} = (\mathbf{e}_i^{(n,l)})^\top \Psi^{(i)}$ *of the subproblems* (4.41), (4.42), *we have* $\|\mathbf{e}_i^{(n,l)}\|_{\mathbf{A}^{(i,i)}} \leq \tilde{R}\cdot\delta$. *Then, the output* $\tilde{e}_i^{(n,l)}$ *from Algorithm 13 fulfills*

$$\|\mathbf{e}_i^{(n,l)} - \tilde{\mathbf{e}}_i^{(n,l)}\|_{\mathbf{A}^{(i,i)}} \leq \delta.$$

Proof. From Equation (4.42), we see that the exact right-hand side of the local subproblems is given by

$$\mathbf{r} := \mathbf{A}^{(i,i)}\mathbf{e}_i^{(n,l)} = (\mathbf{A}\mathbf{w}_i^{(n,l)})_{|\Lambda_i} + \mathbf{f}_{|\Lambda_i} + \mathbf{G}(\mathbf{v}^{(n,l)})_{|\Lambda_i}.$$

In view of the tolerances with which the inexact right-hand side $\tilde{\mathbf{r}}$ is calculated, we therefore have $\|\mathbf{A}^{(i,i)}\mathbf{e}_i^{(n,l)} - \tilde{\mathbf{r}}\|_{\ell_2(\Lambda_i)} \leq 3\,tol$. An application of the Cauchy-Schwarz inequality now gives

$$
\begin{aligned}
\|\mathbf{e}_i^{(n,l)} - (\mathbf{A}^{(i,i)})^{-1}\tilde{\mathbf{r}}\|_{\mathbf{A}^{(i,i)}} &= \left(\langle\mathbf{A}^{(i,i)}\mathbf{e}_i^{(n,l)} - \tilde{\mathbf{r}}, \mathbf{e}_i^{(n,l)} - (\mathbf{A}^{(i,i)})^{-1}\tilde{\mathbf{r}}\rangle_{\ell_2(\Lambda_i)}\right)^{1/2} \\
&= \left(\langle\mathbf{A}^{(i,i)}\mathbf{e}_i^{(n,l)} - \tilde{\mathbf{r}}, (\mathbf{A}^{(i,i)})^{-1}(\mathbf{A}^{(i,i)}\mathbf{e}_i^{(n,l)} - \tilde{\mathbf{r}})\rangle_{\ell_2(\Lambda_i)}\right)^{1/2} \\
&\leq \|(\mathbf{A}^{(i,i)})^{-1}\|_{\ell_2(\Lambda_i)\to\ell_2(\Lambda_i)}^{1/2}\|\mathbf{A}^{(i,i)}\mathbf{e}_i^{(n,l)} - \tilde{\mathbf{r}}\|_{\ell_2(\Lambda_i)} \\
&\leq 3\cdot\|(\mathbf{A}^{(i,i)})^{-1}\|_{\ell_2(\Lambda_i)\to\ell_2(\Lambda_i)}^{1/2}\cdot tol \\
&= \frac{1}{2}\delta.
\end{aligned}
\tag{4.45}
$$

This gives a bound for the initial error by

$$
\begin{aligned}
\|\mathbf{y}^{(0)} - (\mathbf{A}^{(i,i)})^{-1}\tilde{\mathbf{r}}\|_{\mathbf{A}^{(i,i)}} &= \|(\mathbf{A}^{(i,i)})^{-1}\tilde{\mathbf{r}}\|_{\mathbf{A}^{(i,i)}} \\
&\leq \|\mathbf{e}_i^{(n,l)} - (\mathbf{A}^{(i,i)})^{-1}\tilde{\mathbf{r}}\|_{\mathbf{A}^{(i,i)}} + \|\mathbf{e}_i^{(n,l)}\|_{\mathbf{A}^{(i,i)}} \\
&\leq \left(\frac{1}{2} + \tilde{R}\right)\delta.
\end{aligned}
$$

We want to show by induction that

$$
\|\mathbf{y}^{(k)} - (\mathbf{A}^{(i,i)})^{-1}\tilde{\mathbf{r}}\|_{\mathbf{A}^{(i,i)}} \leq \left(\xi^k + \frac{k}{p}\xi^p\right)\cdot\left(\tilde{R} + \frac{1}{2}\right)\delta, \quad 0 \leq k \leq p.
$$

The case $k = 0$ has already been proved. Assume that the assertion holds for $k - 1$. By the tolerance with which $\mathbf{z}^{(k)} := \mathbf{AT\_APPLY}[\mathbf{A}^{(i,i)}, \mathbf{y}^{(k)}, tol_2]$ is computed, we have

$$
\begin{aligned}
\|\mathbf{z}^{(k)} - \mathbf{A}^{(i,i)}\mathbf{y}^{(k)}\|_{\mathbf{A}^{(i,i)}} &\leq \|\mathbf{A}^{(i,i)}\|_{\ell_2(\Lambda_i)\to\ell_2(\Lambda_i)}^{1/2}\|\mathbf{z}^{(k)} - \mathbf{A}^{(i,i)}\mathbf{y}^{(k)}\|_{\ell_2(\Lambda_i)} \\
&\leq \|\mathbf{A}^{(i,i)}\|_{\ell_2(\Lambda_i)\to\ell_2(\Lambda_i)}^{1/2}\cdot tol_2.
\end{aligned}
$$

Since by Equation (4.44), an exact Richardson step gives an error reduction with the factor ξ, using this bound, $\xi < 1$ and the induction hypothesis, we may conclude that

$$
\begin{aligned}
\|\mathbf{y}^{(k)} - (\mathbf{A}^{(i,i)})^{-1}\tilde{\mathbf{r}}\|_{\mathbf{A}^{(i,i)}} &\leq \xi\|\mathbf{y}^{(k-1)} - (\mathbf{A}^{(i,i)})^{-1}\tilde{\mathbf{r}}\|_{\mathbf{A}^{(i,i)}} + \alpha\|\mathbf{A}^{(i,i)}\|_{\ell_2(\Lambda_i)\to\ell_2(\Lambda_i)}^{1/2}\cdot tol_2 \\
&\leq \xi\left(\xi^{k-1} + \frac{k-1}{p}\xi^p\right)\cdot\left(\tilde{R} + \frac{1}{2}\right)\delta + \frac{\xi^p}{p}\left(\tilde{R} + \frac{1}{2}\right)\delta \\
&\leq \left(\xi^k + \frac{k}{p}\xi^p\right)\cdot\left(\tilde{R} + \frac{1}{2}\right)\delta.
\end{aligned}
$$

Thus, with $\tilde{\mathbf{e}}_i^{(n,l)} = \mathbf{y}^{(p)}$, we have $\|\tilde{\mathbf{e}}_i^{(n,l)} - (\mathbf{A}^{(i,i)})^{-1}\tilde{\mathbf{r}}\|_{\mathbf{A}^{(i,i)}} \leq 2\xi^p\left(\tilde{R} + \frac{1}{2}\right)\delta \leq \frac{1}{2}\delta$. Combining this with the estimate (4.45) shows the assertion. $\square$

To sum up, the lemma together with the arguments before shows that *LocSolve* can be applied as a local solver within the multiplicative Schwarz algorithm. Thus, we so far have a convergent, implementable algorithm. In the next steps, we analyze its computational cost and the sparsity of its output.

4.3.4 Optimality of the adaptive method

We are now ready to show that the adaptive multiplicative method is asymptotically optimal. Adapting ideas from [82, 111], we first show that if the exact solution has a representation in $\mathcal{A}_{\mathcal{AT}}^s$, then all the iterates in the algorithm belong to the same approximation class. This will later allow us to establish complexity estimates. Let us start with showing this property for the outer iterates.

Lemma 4.3.7. *Assume that* $\mathbf{P}$ *is bounded on* $\mathcal{A}_{\mathcal{AT}}^s$. *Let* $u = \mathbf{u}^\top \Psi$ *with* $\mathbf{u} \in \mathcal{A}_{\mathcal{AT}}^s$. *Then, the iterates* $\mathbf{u}^{(n)}$ *from Algorithm 12 are also contained in* $\mathcal{A}_{\mathcal{AT}}^s$ *and we have the estimates*

$$\|\mathbf{u}^{(n)}\|_{\mathcal{A}_{\mathcal{AT}}^s} \lesssim \|\mathbf{u}\|_{\mathcal{A}_{\mathcal{AT}}^s},$$
$$\# \operatorname{supp} \mathbf{u}^{(n)} \lesssim \varepsilon_n^{-1/s} \|\mathbf{u}\|_{\mathcal{A}_{\mathcal{AT}}^s}^{1/s}.$$

Proof. The bound (4.40) and Lemma 4.2.3 show that

$$\|\mathbf{u}^{(n)}\|_{\mathcal{A}_{\mathcal{AT}}^s} \lesssim \|\mathbf{P}\mathbf{u}\|_{\mathcal{A}_{\mathcal{AT}}^s}, \quad \# \operatorname{supp} \mathbf{u}^{(n)} \lesssim \varepsilon_n^{-1/s} \|\mathbf{P}\mathbf{u}\|_{\mathcal{A}_{\mathcal{AT}}^s}^{1/s}.$$

Hence, the assertion follows from $\mathbf{P}$ being bounded on $\mathcal{A}_{\mathcal{AT}}^s$. $\qquad\square$

With the help of these estimates, we can now show a similar result even for the inner iterates.

Lemma 4.3.8. *Assume that both* $\mathbf{A}$ *and* $\mathbf{P}$ *are bounded on* $\mathcal{A}_{\mathcal{AT}}^s$. *Then, for the inner iterates* $v^{(n,l)} = (\mathbf{v}^{(n,l)})^\top \Psi$ *and* $w_i^{(n,l)} = (\mathbf{w}_i^{(n,l)})^\top \Psi$, *we have the estimates*

$$\# \operatorname{supp} \mathbf{v}^{(n,l)} \lesssim \varepsilon_n^{-1/s}(\|\mathbf{u}\|_{\mathcal{A}_{\mathcal{AT}}^s}^{1/s} + 1) \tag{4.46}$$

$$\|\mathbf{v}^{(n,l)}\|_{\mathcal{A}_{\mathcal{AT}}^s} \lesssim \|\mathbf{u}\|_{\mathcal{A}_{\mathcal{AT}}^s} + 1 \tag{4.47}$$

and, analogously,

$$\# \operatorname{supp} \mathbf{w}_i^{(n,l)} \lesssim \varepsilon_n^{-1/s}(\|\mathbf{u}\|_{\mathcal{A}_{\mathcal{AT}}^s}^{1/s} + 1) \tag{4.48}$$

$$\|\mathbf{w}_i^{(n,l)}\|_{\mathcal{A}_{\mathcal{AT}}^s} \lesssim \|\mathbf{u}\|_{\mathcal{A}_{\mathcal{AT}}^s} + 1, \tag{4.49}$$

with constants independent of n, l *and* i.

Proof. Let us first assume that $\mathbf{v}^{(n,l)}$ fulfills (4.46) and (4.47). We show by an induction over i that, under this condition, (4.48) and (4.49) are also true.

For $i = 0$, the assertion holds due to $\mathbf{w}_0^{(n,l)} = \mathbf{v}^{(n,l)}$. Let us assume now that the estimates are valid for some $0 \leq i \leq m$. Note that the tolerance δ, with which the local solver *LocSolve* from Subsection 4.3.3 is called, fulfills $\delta \approx \varepsilon_n$. The constants in this relation can be chosen independently of n and l, because $1 \geq \tilde{\rho}^l \geq \tilde{\rho}^{l^*-1}$. Since α and ξ in LocSolve can also be chosen independently from n, l and, if desired, i, we see that the tolerances *tol* and *tol*$_2$ within the local solver are also bounded from below and above by a constant times ε_n. By the properties of the building blocks, see Equations (4.22) for **AT_APPLY**, (4.28) for **EVAL** and (4.31) for **AT_RHS**, for the right-hand side in the local problems, we have the estimate

$$\|\tilde{\mathbf{r}}\|_{\mathcal{A}^s_{\mathcal{AT}}} \lesssim \|\mathbf{v}^{(n,l)}\|_{\mathcal{A}^s_{\mathcal{AT}}} + \|\mathbf{f}\|_{\mathcal{A}^s_{\mathcal{AT}}} + \|\mathbf{v}^{(n,l)}\|_{\mathcal{A}^s_{\mathcal{AT}}} + 1.$$

Here, we have also applied that the restriction of the index set to Λ_i does not increase the $\mathcal{A}^s_{\mathcal{AT}}$-quasi-norm. By assumption, the estimate (4.47) is valid. Moreover, by the boundedness of $\mathbf{A}$ on $\mathcal{A}^s_{\mathcal{AT}}$ and by Theorem 4.2.12, we have

$$\|\mathbf{f}\|_{\mathcal{A}^s_{\mathcal{AT}}} = \|\mathbf{Au} + \mathbf{G}(\mathbf{u})\|_{\mathcal{A}^s_{\mathcal{AT}}} \lesssim \|\mathbf{Au}\|_{\mathcal{A}^s_{\mathcal{AT}}} + \|\mathbf{G}(\mathbf{u})\|_{\mathcal{A}^s_{\mathcal{AT}}} \lesssim \|\mathbf{u}\|_{\mathcal{A}^s_{\mathcal{AT}}} + 1. \qquad (4.50)$$

Thus, it holds that

$$\|\tilde{\mathbf{r}}\|_{\mathcal{A}^s_{\mathcal{AT}}} \lesssim \|\mathbf{u}\|_{\mathcal{A}^s_{\mathcal{AT}}} + 1. \qquad (4.51)$$

In the same fashion, from (4.21), (4.27), (4.31), together with the boundedness of $\mathbf{A}$ on $\mathcal{A}^s_{\mathcal{AT}}$ and Theorem 4.2.12, we have the support estimate

$$\# \operatorname{supp} \tilde{\mathbf{r}} \lesssim \varepsilon_n^{-1/s}(\|\mathbf{u}\|^{1/s}_{\mathcal{A}^s_{\mathcal{AT}}} + 1).$$

Moreover, because $\mathbf{A}$ is s^*-compressible and bounded on $\mathcal{A}^s_{\mathcal{AT}}$, so are its diagonal blocks. Hence, by the estimate (4.22), the inner iterates within *LocSolve* fulfill

$$\|\mathbf{y}^{(r+1)}\|_{\mathcal{A}^s_{\mathcal{AT}}} \lesssim \|\mathbf{y}^{(r)}\|_{\mathcal{A}^s_{\mathcal{AT}}} + \|\tilde{\mathbf{r}}\|_{\mathcal{A}^s_{\mathcal{AT}}}.$$

Combining this with the estimate (4.51) gives

$$\|\mathbf{y}^{(r+1)}\|_{\mathcal{A}^s_{\mathcal{AT}}} \lesssim \|\mathbf{y}^{(r)}\|_{\mathcal{A}^s_{\mathcal{AT}}} + \|\mathbf{u}\|_{\mathcal{A}^s_{\mathcal{AT}}} + 1.$$

Because $\mathbf{y}^{(0)} = 0$ and r runs between 0 and the constant $p - 1$, we see by induction that

$$\|\mathbf{y}^{(r)}\|_{\mathcal{A}^s_{\mathcal{AT}}} \lesssim \|\mathbf{u}\|_{\mathcal{A}^s_{\mathcal{AT}}} + 1 \qquad (4.52)$$

with a constant that can be chosen independently from r. Analogously, by (4.21), we have

$$\# \operatorname{supp} \mathbf{y}^{(r+1)} \lesssim \# \operatorname{supp} \mathbf{y}^{(r)} + \# \operatorname{supp} \tilde{\mathbf{r}} + \varepsilon^{-1/s}\|\mathbf{y}^{(r)}\|^{1/s}_{\mathcal{A}^s_{\mathcal{AT}}},$$

from which we conclude by induction and (4.52) that

$$\# \operatorname{supp} \mathbf{y}^{(r)} \lesssim \varepsilon^{-1/s}(\|\mathbf{u}\|^{1/s}_{\mathcal{A}^s_{\mathcal{AT}}} + 1). \qquad (4.53)$$

In particular, the estimates (4.52) and (4.53) hold for $\tilde{\mathbf{e}}^{(n,l)}_i = \mathbf{y}^{(p)}$. Therefore, going back to Algorithm 12, using this estimate and the induction hypothesis, we see that

$$\|\mathbf{w}^{(n,l)}_{i+1}\|_{\mathcal{A}^s_{\mathcal{AT}}} \lesssim \|\mathbf{w}^{(n,l)}_i\|_{\mathcal{A}^s_{\mathcal{AT}}} + \|\tilde{\mathbf{e}}^{(n,l)}_i\|_{\mathcal{A}^s_{\mathcal{AT}}} \lesssim \|\mathbf{u}\|_{\mathcal{A}^s_{\mathcal{AT}}} + 1,$$

hence the estimate (4.49) is shown. Analogously, by combining the estimate (4.53) with the induction hypothesis, the bound (4.48) for the support of the inner iterates is proved as well. Since i runs between 0 and the constant $m-1$, the implicit constants in these estimates can be chosen independently from n, l and i.

It remains to show that the corresponding estimates for $\mathbf{v}^{(n,l)}$ hold as well, i.e., that (4.46) and (4.47) are valid. For each n, this will be done by an induction over l. Because of $\mathbf{v}^{(n,0)} = \mathbf{u}^{(n)}$, the case $l = 0$ is covered by Lemma 4.3.7. If the assertion is valid for some l, we have seen above that the estimates (4.48) and (4.49) hold. Because $\mathbf{v}^{(n,l+1)} = \mathbf{w}_m^{(n,l)}$, these results directly carry over to $\mathbf{v}^{(n,l+1)}$. Moreover, because l is between 0 and the constant number l^*, the implicit constants in the above estimates do not mount up uncontrollably, but are bounded by a global constant independent of n and l. Thus, the proof is completed. □

The preceding results have shown that our method is asymptotically optimal with respect to the degrees of freedom of the output and the inner iterates. It remains to show an analogous result for the computational complexity. The following lemma will help us to do so.

Lemma 4.3.9. *Under the assumptions from Lemma 4.3.8, the computation of $\mathbf{v}^{(n,l+1)}$ from $\mathbf{v}^{(n,l)}$ requires at most a constant multiple of $\varepsilon_n^{-1/s}(\|\mathbf{u}\|_{\mathcal{A}_{\mathcal{AT}}^s}^{1/s} + 1)$ operations.*

Proof. The computational complexity of the calculation basically amounts to m calls of *LocSolve*. The additional calculation of $\mathbf{w}_{i+1}^{(n,l)} = \mathbf{w}_i^{(n,l)} + \tilde{\mathbf{e}}_i^{(n,l)}$ can of course be done in at most $\#\operatorname{supp} \mathbf{w}_i^{(n,l)} + \#\operatorname{supp} \tilde{\mathbf{e}}_i^{(n,l)}$ operations, which by (4.48) and (4.53) is bounded by a constant multiple of $\varepsilon_n^{-1/s}(\|\mathbf{u}\|_{\mathcal{A}_{\mathcal{AT}}^s}^{1/s} + 1)$. Since m is constant, it is sufficient to analyze the number of iterations within one call of *LocSolve*. By Proposition 4.2.7, Proposition 4.2.15 and the assumption from Subsection 4.2.4 on the computational complexity of **AT_RHS**, the calculation of the right-hand side $\tilde{\mathbf{r}}$ requires at most a constant multiple of

$$\varepsilon_n^{-1/s}\|\mathbf{w}_i^{(n,l)}\|_{\mathcal{A}_{\mathcal{AT}}^s}^{1/s} + \#\mathcal{T}(\operatorname{supp} \mathbf{w}_i^{(n,l)})$$
$$+ \varepsilon_n^{-1/s}\|\mathbf{f}\|_{\mathcal{A}_{\mathcal{AT}}^s}^{1/s}$$
$$+ \varepsilon_n^{-1/s}(\|\mathbf{v}^{(n,l)}\|_{\mathcal{A}_{\mathcal{AT}}^s}^{1/s} + 1) + \#\mathcal{T}(\operatorname{supp} \mathbf{v}^{(n,l)})$$

operations. Since $\mathbf{w}_i^{(n,l)}$ and $\mathbf{v}^{(n,l)}$ already are in tree structure, by Lemma 4.3.8 and the estimate (4.50), this expression can in turn be bounded by a constant times $\varepsilon_n^{-1/s}(\|\mathbf{u}\|_{\mathcal{A}_{\mathcal{AT}}^s}^{1/s} + 1)$. As before, the work needed for the addition of the vectors can be neglected, because their support sizes are of the same order, compare the proof of Lemma 4.3.8. By Proposition 4.2.7 and $tol_2 \approx \varepsilon_n$, the calculation of $\mathbf{z}^{(r)} :=$ **AT_APPLY**$[\mathbf{A}^{(i,i)}, \mathbf{y}^{(r)}, tol_2]$ requires at most a constant times

$$\varepsilon_n^{-1/s}\|\mathbf{y}^{(r)}\|_{\mathcal{A}_{\mathcal{AT}}^s}^{1/s} + \#\mathcal{T}(\operatorname{supp} \mathbf{y}^{(r)})$$

operations. Since $\mathbf{y}^{(r)}$ is already tree-structured, by the estimates (4.52) and (4.53), this term can be bounded by a constant times $\varepsilon_n^{-1/s}(\|\mathbf{u}\|_{\mathcal{A}_{\mathcal{AT}}^s}^{1/s} + 1)$. Hence, in view of the bounds for the support of $\mathbf{y}^{(r)}$ and $\tilde{\mathbf{r}}$, the costs for computing $\mathbf{y}^{(r+1)}$ from $\mathbf{y}^{(r)}$ fulfill an analogous estimate. Since p, and thus the number of inner iterations within *LocSolve*, is constant, the overall cost for one call of the local solver is at most a constant multiple of $\varepsilon_n^{-1/s}(\|\mathbf{u}\|_{\mathcal{A}_{\mathcal{A}\mathcal{T}}^s}^{1/s} + 1)$. Thus, the assertion is shown. $\qquad\square$

By summarizing the previous results, we are now ready to state the first main result of this thesis. This result shows that the adaptive multiplicative Schwarz method is convergent and asymptotically optimal.

Theorem 4.3.10. *Let $0 < s < s^*$, $\mathbf{A}$ and $\mathbf{P}$ be s^*-compressible and bounded on $\mathcal{A}_{\mathcal{A}\mathcal{T}}^s$. Let $\rho < 1$ and $u = \mathbf{u}^\top\Psi$ be the exact solution of (4.1). Then, for any $\varepsilon > 0$, the output $u_\varepsilon = \mathbf{u}_\varepsilon^\top\Psi$ from the adaptive multiplicative Schwarz method fulfills*

$$\|u_\varepsilon - u\| \le \varepsilon. \tag{4.54}$$

If, in addition, $\mathbf{u} \in \mathcal{A}_{\mathcal{A}\mathcal{T}}^s$, then we have

$$\|\mathbf{u}_\varepsilon\|_{\mathcal{A}_{\mathcal{A}\mathcal{T}}^s} \lesssim \|\mathbf{u}\|_{\mathcal{A}_{\mathcal{A}\mathcal{T}}^s}, \quad \#\operatorname{supp}\mathbf{u}_\varepsilon \lesssim \varepsilon^{-1/s}(\|\mathbf{u}\|_{\mathcal{A}_{\mathcal{A}\mathcal{T}}^s}^{1/s} + 1). \tag{4.55}$$

Moreover, the number of operations needed to compute $\mathbf{u}_\varepsilon$ is bounded by a constant times $\varepsilon^{-1/s}(\|\mathbf{u}\|_{\mathcal{A}_{\mathcal{A}\mathcal{T}}^s}^{1/s} + 1)$.

Proof. The convergence result (4.54) has already been shown in Proposition 4.3.3. Let $N \in \mathbb{N}$ be minimal so that $\varepsilon_N \le \varepsilon$, hence Algorithm 12 terminates with $\mathbf{u}_\varepsilon = \mathbf{u}^{(N)}$. Because the ε_n are geometrically decreasing, it is $\varepsilon_N \eqsim \varepsilon$. Thus, (4.55) follows from Lemma 4.3.7.

Let us now discuss the computational complexity. Since the number l^* is constant, by Lemma 4.3.9 the computation of $\mathbf{v}^{(n,l^*)}$ from $\mathbf{u}^{(n)}$ needs a constant times $\varepsilon_n^{-1/s}(\|\mathbf{u}\|_{\mathcal{A}_{\mathcal{A}\mathcal{T}}^s}^{1/s} + 1)$ operations. By Proposition 4.2.7 combined with Lemma 4.3.8 and the fact that the $\mathbf{v}^{(n,l)}$ are tree-structured, the computation of $\tilde{\mathbf{u}}^{(n+1)}$ from $\mathbf{v}^{(n,l)}$ by the approximate application of $\mathbf{P}$ requires a constant multiple of

$$\left(\frac{\varepsilon_{n+1}}{2K_1(2C^* + 1)}\right)^{-1/s} \|\mathbf{v}^{(n,l)}\|_{\mathcal{A}_{\mathcal{A}\mathcal{T}}^s}^{1/s} + \#\mathcal{T}(\operatorname{supp}\mathbf{v}^{(n,l)}) \lesssim \varepsilon_{n+1}^{-1/s}(\|\mathbf{u}\|_{\mathcal{A}_{\mathcal{A}\mathcal{T}}^s}^{1/s} + 1)$$

operations and $\#\operatorname{supp}\tilde{\mathbf{u}}^{(n+1)}$ fulfills an analogous estimate. Hence, the final application of **AT_COARSE** to compute $\mathbf{u}^{(n+1)}$ from the tree-structured vector $\tilde{\mathbf{u}}^{(n+1)}$ also requires not more than of the order $\varepsilon_{n+1}^{-1/s}(\|\mathbf{u}\|_{\mathcal{A}_{\mathcal{A}\mathcal{T}}^s}^{1/s} + 1)$ operations. Altogether, the total number of operations needed to compute $\mathbf{u}_\varepsilon$ is therefore bounded by a constant times

$$\sum_{n=1}^{N} \varepsilon_n^{-1/s}(\|\mathbf{u}\|_{\mathcal{A}_{\mathcal{A}\mathcal{T}}^s}^{1/s} + 1).$$

By a standard geometric series argument, because of $\varepsilon_n = \varepsilon_N\tilde{\rho}^{n-N}$ and $\tilde{\rho}^{1/s} < 1$, we have

$$\sum_{n=1}^{N} \varepsilon_n^{-1/s} = \varepsilon_N^{-1/s}\sum_{n=1}^{N}(\tilde{\rho}^{1/s})^{N-n} \le \varepsilon_N^{-1/s}\sum_{k=0}^{\infty}(\tilde{\rho}^{1/s})^k \lesssim \varepsilon_N^{-1/s} \lesssim \varepsilon^{-1/s},$$

which completes the proof. $\hfill\square$

4.4 Adaptive additive wavelet Schwarz method

In the following, we formulate an adaptive wavelet version of the *additive* Schwarz method for nonlinear problems. As previously outlined, the main advantage of this version is that it allows for a parallel solution of the subproblems. The methods and the results presented in this Section 4.4 have, in similar form, already been published in [82].

4.4.1 The adaptive method and its convergence

With the notation and the constants already introduced in Subsections 4.1.2 and 4.3.2, we are now ready to formulate the adaptive wavelet version of the additive Schwarz method from Algorithm 4. The solution of the subproblems will again be described afterwards.

Algorithm 14 AddSchw3 $[\varepsilon]$

$\quad$ % *Let* $M \geq \|u\|$.
$\quad$ % *Let* C^* *be the constant in Lemma 4.2.3.*
$\quad$ % *Let* $l^* \in \mathbb{N}$ *be minimal with* $\tilde{\varrho}^{l^*} \leq \frac{1}{2K_1 K_2} \frac{1}{2C^*+1} \tilde{\varrho}$.
$\quad$ % *Let* $\varepsilon_n := \tilde{\varrho}^n M$, $n \in \mathbb{N}$.
$\quad u^{(0)} := 0$
$\quad n := 0$
$\quad$ **while** $\varepsilon_n > \varepsilon$ **do**
$\quad\quad v^{(n,0)} := u^{(n)}$
$\quad\quad$ **for** $l = 0, \ldots, l^* - 1$ **do**
$\quad\quad\quad$ **for** $i = 0, \ldots, m - 1$ **do**
$\quad\quad\quad\quad$ Compute $\tilde{e}_i^{(n,l)} \in H_0^t(\Omega_i)$ as an approximation to $e_i^{(n,l)} \in H_0^t(\Omega_i)$ from
$\quad\quad\quad\quad a(e_i^{(n,l)}, v) = -a(v^{(n,l)}, v) + f(v) - \mathcal{G}(v^{(n,l)})(v), \quad v \in H_0^t(\Omega_i)$
$\quad\quad\quad\quad$ with tolerance $\|e_i^{(n,l)} - \tilde{e}_i^{(n,l)}\| \leq \frac{1}{2m\omega}\varepsilon_n \tilde{\varrho}^l(1-\varrho)$.
$\quad\quad\quad$ **end for**
$\quad\quad\quad v^{(n,l+1)} := v^{(n,l)} + \omega \sum_{i=0}^{m-1} \tilde{e}_i^{(n,l)}$
$\quad\quad$ **end for**
$\quad\quad \tilde{\mathbf{u}}^{(n+1)} := \mathbf{AT\_APPLY}[\mathbf{P}, \mathbf{v}^{(n,l^*)}, \frac{1}{2K_1} \frac{1}{2C^*+1}\varepsilon_{n+1}]$
$\quad\quad \mathbf{u}^{(n+1)} := \mathbf{AT\_COARSE}[\tilde{\mathbf{u}}^{(n+1)}, \frac{1}{K_1} \frac{2C^*}{2C^*+1}\varepsilon_{n+1}]$
$\quad\quad n := n + 1$
$\quad$ **end while**
$\quad \mathbf{u}_\varepsilon := \mathbf{u}^{(n)}$
$\quad u_\varepsilon := \mathbf{u}_\varepsilon^\top \Psi$

Taking a similar approach in the analysis of this method as in the previous section, we first establish the convergence of this algorithm.

Proposition 4.4.1. *Let $\varrho < 1$. Then, the iterates $u^{(n)}$ from Algorithm 14 fulfill*

$$\|u^{(n)} - u\| \leq \varepsilon_n.$$

Hence, the method terminates after finitely many iterations with $\|u_\varepsilon - u\| \leq \varepsilon$.

Proof. We use induction over n, similarly to the proof of Proposition 4.3.3. By choice of M and $u^{(0)}$, the case $n = 0$ is clear. Assume that the assertion holds for some $n \in \mathbb{N}$. As in the proof of Proposition 4.1.8, by induction we have

$$\|v^{(n,l)} - u\| \leq \tilde{\varrho}^l \varepsilon_n. \tag{4.56}$$

In particular, by the choice of l^* in the algorithm, K_2 from (4.38) and the tolerance for the call of **AT_APPLY**, it follows that

$$\begin{aligned}
\|\tilde{\mathbf{u}}^{(n+1)} - \mathbf{Pu}\|_{\ell_2(\Lambda)} &\leq \|\tilde{\mathbf{u}}^{(n+1)} - \mathbf{Pv}^{(n,l^*)}\|_{\ell_2(\Lambda)} + K_2 \|v^{(n,l^*)} - u\| \\
&\leq \frac{1}{2K_1} \frac{1}{2C^* + 1} \varepsilon_{n+1} + K_2 \frac{1}{2K_1 K_2} \frac{1}{2C^* + 1} \tilde{\varrho} \varepsilon_n \\
&= \frac{1}{K_1} \frac{1}{2C^* + 1} \varepsilon_{n+1}.
\end{aligned} \tag{4.57}$$

Thus, with the given tolerance for the coarsening step and the choice of K_1 from (4.37), we have

$$\begin{aligned}
\|u^{(n+1)} - u\| &\leq K_1 \left(\|\tilde{\mathbf{u}}^{(n+1)} - \mathbf{Pu}\|_{\ell_2(\Lambda)} + \|\tilde{\mathbf{u}}^{(n+1)} - \mathbf{u}^{(n+1)}\|_{\ell_2(\Lambda)} \right) \\
&\leq \frac{1}{2C^* + 1} \varepsilon_{n+1} + \frac{2C^*}{2C^* + 1} \varepsilon_{n+1} \\
&= \varepsilon_{n+1},
\end{aligned}$$

and the proof is completed. $\qquad\square$

4.4.2 Solution of the local subproblems

To a large extent, the solution of the local subproblems can be tackled following the lines from Subsection 4.3.3. Doing so, analogous complexity and sparsity estimates can be shown, as we will explain in a little more detail. The local subproblems now assume the form

$$a(e_i^{(n,l)}, v) = -a(v^{(n,l)}, v) + f(v) - \mathcal{G}(v^{(n,l)})(v), \quad v \in H_0^t(\Omega_i), \tag{4.58}$$

and the equivalent discrete problems are given by

$$\mathbf{A}^{(i,i)} \mathbf{e}_i^{(n,l)} = -(\mathbf{Av}^{(n,l)})_{|\Lambda_i} + \mathbf{f}_{|\Lambda_i} - \mathbf{G}(\mathbf{v}^{(n,l)})_{|\Lambda_i}. \tag{4.59}$$

A central step in the analysis of the local solvers in the multiplicative Schwarz method was Lemma 4.3.4, which had the consequence that only a fixed number of iterations was needed within each call of *LocSolve*. To show corresponding results for the additive method, we need an analogous estimate for the solution of the problems (4.58).

Lemma 4.4.2. *The exact solutions $e_i^{(n,l)}$ of the local subproblems in Algorithm 14 fulfill the estimate $\|e_i^{(n,l)}\| \lesssim \varepsilon_n \tilde{\varrho}^l$ with a constant independent of n, i and l.*

Proof. Analogous to Equation (4.17), the local correction now has the representation

$$e_i^{(n,l)} = P_i(u - v^{(n,l)}) - \mathcal{L}_i^{-1}(\mathcal{G}(v^{(n,l)}) - \mathcal{G}(u)).$$

Using that $\|P_i\| = 1$, that $\mathcal{L}_i$ is an isometry with respect to the energy norm, the contraction property of $\mathcal{G}$ and the estimate (4.56), we end up with

$$\begin{aligned}
\|e_i^{(n,l)}\| &\leq \|P_i(u - v^{(n,l)})\| + \|\mathcal{L}_i^{-1}(\mathcal{G}(v^{(n,l)}) - \mathcal{G}(u))\| \\
&\leq (1+c)\|v^{(n,l)} - u\| \\
&\leq (1+c)\tilde{\varrho}^l \varepsilon_n.
\end{aligned}$$

$\square$

With this lemma at hand, the application of the local solvers carries over nearly verbatim. Only the computation of the approximate discrete right-hand side has to be adapted, so we replace this line in *LocSolve* by

$$\tilde{\mathbf{r}} := -\mathbf{AT\_APPLY}[\mathbf{A}, \mathbf{v}^{(n,l)}, tol]_{|\Lambda_i} + \mathbf{AT\_RHS}[\mathbf{f}, tol]_{|\Lambda_i} - \mathbf{EVAL}[\mathbf{G}, \mathbf{v}^{(n,l)}, tol]_{|\Lambda_i}.$$

Obviously, the convergence of the local solvers is unaffected by this modification, so Lemma 4.3.6 remains valid.

4.4.3 Optimality of the adaptive method

We are now ready to show the second main result of this chapter, the convergence and optimality of the additive Schwarz method. The proof is done in a similar way as for the multiplicative algorithm.

Theorem 4.4.3. *Let $0 < s < s^*$, assume that $\mathbf{A}$ and $\mathbf{P}$ are s^*-compressible and bounded on $\mathcal{A}_{AT}^s$. Let $\varrho < 1$ and $u = \mathbf{u}^\top \Psi$ be the exact solution of (4.1). Then, for any $\varepsilon > 0$, the output $u_\varepsilon = \mathbf{u}_\varepsilon^\top \Psi$ from Algorithm 14 with the local solver described in Subsection 4.4.2 fulfills*

$$\|u_\varepsilon - u\| \leq \varepsilon. \tag{4.60}$$

If, in addition, $\mathbf{u} \in \mathcal{A}_{AT}^s$, then we have

$$\|\mathbf{u}_\varepsilon\|_{\mathcal{A}_{AT}^s} \lesssim \|\mathbf{u}\|_{\mathcal{A}_{AT}^s}, \quad \# \mathrm{supp}\, \mathbf{u}_\varepsilon \lesssim \varepsilon^{-1/s}(\|\mathbf{u}\|_{\mathcal{A}_{AT}^s}^{1/s} + 1). \tag{4.61}$$

Moreover, the number of operations needed to compute $\mathbf{u}_\varepsilon$ is bounded by a constant times $\varepsilon^{-1/s}(\|\mathbf{u}\|_{\mathcal{A}_{AT}^s}^{1/s} + 1)$.

Proof. The convergence result (4.60) has already been shown in Proposition 4.4.1 together with the convergence of the local solver. The optimality results are verified in a very similar fashion as in Theorem 4.3.10, so we give a brief presentation of the main lines. First of all, from the error bound (4.57), the Coarsening Lemma 4.2.3 and the boundedness assumption on $\mathbf{P}$, we see analogously to Lemma 4.3.7 that $\mathbf{u}^{(n)} \in \mathcal{A}_{\mathcal{A}\mathcal{T}}^s$ with $\|\mathbf{u}^{(n)}\|_{\mathcal{A}_{\mathcal{A}\mathcal{T}}^s} \lesssim \|\mathbf{u}\|_{\mathcal{A}_{\mathcal{A}\mathcal{T}}^s}$ and

$$\# \operatorname{supp} \mathbf{u}^{(n)} \lesssim \varepsilon_n^{-1/s} \|\mathbf{u}\|_{\mathcal{A}_{\mathcal{A}\mathcal{T}}^s}^{1/s},$$

uniformly in $n \in \mathbb{N}$. This already shows (4.61). Using the properties of the building blocks **AT_APPLY**, **EVAL** and **AT_RHS**, an inductive argument as in the proof of Lemma 4.3.8 shows that $\mathbf{v}^{(n,l)} \in \mathcal{A}_{\mathcal{A}\mathcal{T}}^s$ with $\|\mathbf{v}^{(n,l)}\|_{\mathcal{A}_{\mathcal{A}\mathcal{T}}^s} \lesssim \|\mathbf{u}\|_{\mathcal{A}_{\mathcal{A}\mathcal{T}}^s} + 1$ and

$$\# \operatorname{supp} \mathbf{v}^{(n,l)} \lesssim \varepsilon_n^{-1/s} (\|\mathbf{u}\|_{\mathcal{A}_{\mathcal{A}\mathcal{T}}^s}^{1/s} + 1).$$

with constants that are uniformly bounded in n and l. Moreover, with the same arguments as in Lemma 4.3.9, the effort for computation $\mathbf{v}^{(n,l+1)}$ from $\mathbf{v}^{(n,l)}$ is bounded by a constant times $\varepsilon_n^{-1/s} (\|\mathbf{u}\|_{\mathcal{A}_{\mathcal{A}\mathcal{T}}^s}^{1/s} + 1)$ operations. Together with the complexity estimates for **AT_APPLY** and **AT_COARSE**, and the fact that $\sum_{\varepsilon_n \geq \varepsilon} \varepsilon_n^{-1/s} \lesssim \varepsilon^{-1/s}$, this allows us to estimate the overall complexity of the algorithm by a constant multiple of $\varepsilon^{-1/s} (\|\mathbf{u}\|_{\mathcal{A}_{\mathcal{A}\mathcal{T}}^s}^{1/s} + 1)$ as in the proof of Theorem 4.3.10. $\qquad\square$

Chapter 5

Numerical tests

In this chapter, we test the adaptive algorithms on representative model problems in one and two space dimensions. The aim of these experiments is to verify the convergence and optimality results derived in the previous chapter. In particular, we will see that the expected asymptotic rates can be observed in practice at realistic scales. To make the results comparable, we run the algorithms on standard problems, see, e.g., [33, 35, 75, 76, 111, 123]. These problems are well-suited for testing purposes, because they cover important criteria for test of adaptive methods such as the resolution of singularities induced by the right-hand side and by the domain. The results presented here extend and amplify the experiments from [82], albeit with a slightly different choice of the involved constants.

5.1 The practical implementation of the methods

Let us first make some general remarks on the implementation of the methods. These remarks are valid for both the multiplicative and the additive method. The algorithms are realized in $C++$ with tools and building blocks developed at the Numerics and Optimization Group at Philipps-Universität Marburg. Large parts of the underlying code have been published in the *Wavelet and Multiscale Library* that is available at [1]. In particular, this library implements the spline wavelet construction from [95], that will be applied in the experiments. For plotting the results, we will, in addition, make use of tools from *Matlab*.

As we will explain now, in the practical realization of the methods, a few changes are made compared to the algorithms defined in Sections 4.3 and 4.4, compare also [76, 97, 123]. Let us first note that many constants involved in the calculation of the tolerances are usually not known explicitly, a typical example are the constants that appear in norm equivalences. Hence, they are replaced with estimates or empirically derived fixed values. In addition, it has turned out that data structures are easier to handle if we only consider wavelets up to a fixed maximal level $j_{\max}$. This also allows us to precompute the right-hand side and the important diagonal elements of the stiffness matrix up to this level. However, in principle we expect that it is possible to implement the algorithms without this limitation. This, however, would require some modifications of the library [1]. Moreover, as already outlined in Subsection 4.3.1, the explicit application of the projector **P** helps to theoretically verify the optimality of the algorithm, but is less desirable from a practical point of view. Therefore, it is left out in the numerical experiments. Another observation made in numerical tests is that the support size of the inner iterates can become undesirably large between two calls of the coarsening step. To avoid this, we call this method more often than theoretically necessary. For instance, an additional coarsening step is introduced after each inner iteration within *MultSchw3* and *AddSchw3*, and not only after l^* such iterations. Because it is much faster in practice, for this step we make use of a classical coarsening method as in [25] although it does not necessarily generate a complete tree, see also Subsection 4.2.1.

In addition, we make use of some practical improvements of the method **EVAL**, that were developed in [76]. Let us shortly explain these modifications. Recall that an important step in this routine is the determination of the support of the output via the subroutine **SUPPORT_PREDICTION**. In this step, we repeatedly have to check whether or not a pair of wavelets has intersecting support. Because this calculation is costly, it has turned out that it pays off to save the result of this calculation in a list, so that the information can in later steps be accessed without further computations. In a similar way, point evaluations of wavelets are stored.

5.2 Numerical tests on the interval

At first, we apply the algorithms to the standard test problem

$$-\Delta u + u^3 = f \text{ in } \mathcal{I} = (0,1), \quad u(0) = u(1) = 0$$

from Example 1.3.1 on the unit interval with the exact solution

$$u(x) := -\sin(3\pi x) + \begin{cases} 2x^2, & x \in (0, \frac{1}{2}) \\ 2(1-x)^2, & x \in [\frac{1}{2}, 1) \end{cases}.$$

The corresponding right-hand side f is given by $f(v) = 4v(\frac{1}{2}) + \int\limits_0^1 g(x)v(x)\,\mathrm{d}x$, where

$$g(x) = -9\pi^2 \sin(3\pi x) - 4 + u(x)^3,$$

see [75, Section 5.1].

The solution u is chosen as a function with a point singularity at $x = 0.5$ induced by the right-hand side. It can be shown, see, e.g., [123, Section 4.4.1], that this singularity limits the Sobolev regularity, so that u belongs to $H^\alpha(\Omega)$ only for $\alpha < \frac{3}{2}$. Hence, in view of the results from Subsection 3.2.1, we expect at most a rate of $s^* = \frac{1}{2}$ from standard uniform methods. On the other hand, the solution belongs to all Besov spaces $B^{s+1}_{\tau,\tau}(\Omega)$, $\frac{1}{\tau} = s + \frac{1}{2}$. Using interpolation, see Proposition 3.1.2, we can also conclude that it is contained in all the spaces in the slightly stronger scale $B^{s+1}_{\tau,\tau}(\Omega)$,

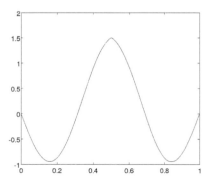

Figure 5.1: The exact solution u of the one-dimensional test problem.

$\frac{1}{\tilde{\tau}} < \frac{1}{\tau}$ required in Proposition 3.4.7. Hence, we can expect the solution to have a discrete representation in $\mathcal{A}^s_{\mathcal{AT}}$ for all $s < s^* = l - 1$. Thus, we expect a convergence rate of $(l-1)$ from an asymptotically optimal adaptive algorithm. In particular, in this case the rate is only limited by the order of the underlying wavelet basis. By the above arguments, this example should be suitable to demonstrate the benefits of using adaptivity to detect the singularity and exploit the unlimited regularity of the function in the adaptivity scale compared to its very limited Sobolev regularity.

To construct an aggregated wavelet frame, we fix a decomposition of the unit interval by setting $\Omega_0 := (0, 0.7)$ and $\Omega_1 := (0.3, 1)$. This domain decomposition is overlapping in the sense of Definition 2.2.8, see Example 2.2.14. On the subdomains Ω_0 and Ω_1, we use the wavelet basis from [95] with primal and dual orders l and $\tilde{l}$, chosen from $(l, \tilde{l}) \in \{(2,2), (3,3), (4,6)\}$. This corresponds to piecewise linear, quadratic and cubic spline wavelets, respectively. We set the maximum level of the wavelets considered to $j_{\max} := 15$. The minimum level depends on the order of the basis and is given by $j_0 = 3$ for $(l, \tilde{l}) \in \{(2,2), (3,3)\}$ and $j_0 = 5$ for $(l, \tilde{l}) = (4,6)$.

As choice of parameters, we set $\rho := \varrho := 0.6$ as the expected error reduction of the exact multiplicative and additive Schwarz method. Moreover, the parameter c from (1.3) is set to $c := 0.5$. These values were chosen empirically, which means that although we cannot prove that these are correct choices in theory, the algorithms perform well with this set of parameters. In view of Proposition 4.1.7, the relaxation parameter of the additive Schwarz method is set to $\omega := \frac{1}{m} = 0.5$. Moreover, we set $K_1 := K_2 := C^* := 1$ for the unknown constants in the norm equivalences and the coarsening method. In the local Richardson solver within *LocSolve*, we make 50 iterations with a relaxation parameter $\alpha := 0.25$ and set $\xi := 0.8$.

For testing purposes, we fix the number of outer iterations within *MultSchw3* and *AddSchw3* at $n_{\max} := 50$. As a proxy for the error, we use an approximate discrete residual that is calculated using the previous iterate. This allows us to reuse the expensive evaluation of the discrete nonlinearity that is calculated anyway when setting up the right-hand side vector $\tilde{\mathbf{r}}$ within *LocSolve*.

On the left-hand side in Figure 5.2, the number of degrees of freedom of the iterates $\mathbf{u}^{(n)}$ within *MultSchw3* and *AddSchw3* is compared to the approximate residual. We see that the optimal rate, which corresponds to the slope indicated in the graph with a line, is nearly reproduced by both the multiplicative and the additive Schwarz method. The multiplicative method tends to give slightly better results. This is not completely unexpected, because in the multiplicative variant, contrary to the additive method, the information from the local solver on Ω_0 is used for the solution on Ω_1. For piecewise linear wavelets, i.e., in the case $(l, \tilde{l}) = (2, 2)$, we see that the support sizes become inconveniently large and end up at close to 10^5. The test with piecewise quadratic wavelets, $(l, \tilde{l}) = (3, 3)$, shows a relatively steady decay of the residual at a rate very close to the expected rate. In the tests with piecewise cubic wavelets, $(l, \tilde{l}) = (4, 6)$, we see that, initially, the support sizes of the iterates grow very fast. After this initial phase, however, the residual decays rapidly. In all three cases, the rate tends to become a little smaller than expected towards the end of the iteration. This might partly be due to employing wavelets up to the maximum level $j_{\max}$, where further refinement is not possible.

In a similar fashion, on the right-hand side in Figure 5.2, we compare the computational time to the approximate residual. The time was measured on a Lenovo ThinkPad with four CPUs at 2.2 GHz and the work for precomputations and the preparation of the plots is excluded. In absolute terms, the algorithm ran fastest for the basis with parameters $(l, \tilde{l}) = (3, 3)$. There, we were also able to observe the expected rate at least for some time after the initial phase. For the other two configurations, we observe a rate close to, but slightly below, the expected rate during large parts of the running time. It appears from the experiments that the main bottleneck regarding the computational effort lies in the evaluation of the nonlinear term, which poses quite a few practical difficulties, compare also the experiments in [76].

In Figure 5.3, we compare the multiplicative adaptive Schwarz method to a simple uniform wavelet method. This uniform Galerkin-type method computes approximations $\mathbf{u}^{(j)}$ using wavelets only up to level j, i.e., by projecting the equation onto V_j, compare Subsection 3.2.1. The resulting finite-dimensional systems are solved approximately up to a tolerance of $5 \cdot 10^{-4}$ with a Richardson iteration acting on the full index set up to level j. The approximation is then gradually refined by increasing j stepwise from $j = 4$ to $j = 12$. For the convergence history, the residual is approximated with $j_{\max} = 15$ up to a very small tolerance. The test is performed with piecewise quadratic wavelets, $(l, \tilde{l}) = (3, 3)$. We see in the graph that this method realizes the maximal rate $s^* = \frac{1}{2}$ that we can expect from such a standard uniform method for a solution contained in $H^{s+1}(\Omega)$ for $s < s^* = \frac{1}{2}$, compare the discussion at the beginning of this section. In comparison, the adaptive multiplicative Schwarz method not only realizes a higher rate $s^* = \frac{l-t}{d} = 2$, but also gives sparser approximations in absolute terms from the very beginning of the iteration. Hence, in this case,

$(l, \tilde{l}) = (2, 2):$

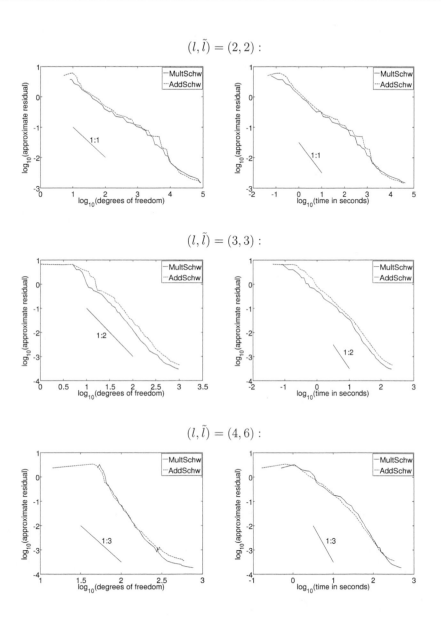

$(l, \tilde{l}) = (3, 3):$

$(l, \tilde{l}) = (4, 6):$

Figure 5.2: Degrees of freedom vs. ℓ_2-norm of the approximate residual (left) and computational time vs. ℓ_2-norm of the approximate residual (right) for the one-dimensional test problem.

the adaptive approach not only outperforms this uniform method in the asymptotic regime, but already in early stages. However, it has to be noted that the pointwise error of the uniform and the adaptive method are of a similar magnitude at the end of the iteration although the residual for the uniform method is larger. This might be due to the fact that L_∞-norm of the wavelets in $H_0^1(\mathcal{I})$ behaves like $2^{-j/2}$, so leaving away wavelets on high levels has a smaller impact on the pointwise error than on the error in $H^1(\mathcal{I})$ or on the residual.

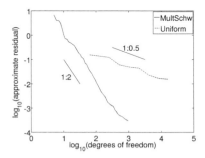

Figure 5.3: Degrees of freedom vs. ℓ_2-norm of the approximate residual of the adaptive multiplicative Schwarz and a uniform method in the one-dimensional test case with $(l, \tilde{l}) = (3, 3)$.

In Figure 5.4, we see the pointwise error after the maximum number $n_{\max} = 50$ of iterations from the multiplicative Schwarz algorithm on the left. In the center and on the right, the local approximations from the subdomains are shown. Let us start with discussing the pointwise error. Although it tends to be largest around the singularity at $x = 0.5$, the behavior of the error is still relatively even on the whole domain and the singularity is resolved quite well. For all three basis parameters, we end up with a final maximum pointwise error $\|u_\varepsilon - u\|_{L_\infty(\mathcal{I})}$ in the order of 10^{-5}.

Let us now explain the right-hand side in Figure 5.4. Since we have left out the explicit application of the projector $\mathbf{P}$ in the implementation, it is not immediately clear how the approximate solution $u_\varepsilon = \mathbf{u}_\varepsilon^\top \Psi \in H_0^t(\Omega)$ from the algorithm decomposes into its local approximations $u_i = (\mathbf{u}_\varepsilon)_{|\Lambda_i}^\top \Psi^{(i)} \in H_0^t(\Omega_i)$, $i = 0, 1$. In particular, we are interested in the behavior of the local approximations in the overlapping region $\Omega_0 \cap \Omega_1$. Oscillations in this area might point at redundancies. Although they cancel out in the overall approximation $u_\varepsilon = u_0 + u_1$, such oscillations would lead to additional computational costs. Fortunately, in the experiments we observe very few such oscillations. In particular, the representation calculated in the case $(l, \tilde{l}) = (3, 3)$ seems economic in the sense that the representation of the approximate solution in the overlapping region is almost completely done by the local approximation on Ω_0.

In Figure 5.5, we see the pointwise error and the local approximations for the additive Schwarz method. For the bases with parameters $(l, \tilde{l}) \in \{(2,2), (3,3)\}$, we end up with a maximum pointwise error of about $\|u_\varepsilon - u\|_{L_\infty(\mathcal{I})} \approx 6 \cdot 10^{-5}$. For the basis $(l, \tilde{l}) = (4,6)$, the error is about one magnitude larger, which is somewhat surprising but might partly correspond to the behavior of the residual observed in Figure 5.2. Similarly to the multiplicative Schwarz method, the local approximations show very few oscillations or obvious redundancies. It is striking that the local approximations are more symmetric than in the multiplicative method. This is not surprising, but reflects the more symmetric character of the additive method itself.

Furthermore, in Figure 5.6, we see the distribution of the active wavelet coefficients on the subintervals Ω_0, Ω_1 during the course of the iteration. As a representative example, we consider the additive Schwarz method with $(l, \tilde{l}) = (3,3)$. The graphs, in particular the situations after 20 and 30 iterations, show very well that the adaptive method detects the singularity of the solution at $x = 0.5$ and refines the approximation there by adding additional wavelet on higher scales. After around 50 iterations, we have to observe some redundancies in the overlapping region. This contrasts somewhat with Figure 5.5, because the splitting into the local parts seems economic at first sight, but there might be a few redundancies on higher scales. In principle, these redundancies can be controlled by the implicit application of the projector $\mathbf{P}$ as described in Section 4.3.1. However, the experiments made in [81] in the context of adaptive methods for linear problems suggest that the overall performance of the algorithm is not improved by this modification. This is partly because the coordination between the projection step and the other parts of the algorithm may be difficult in practice because many of the constants involved in the parameters are not explicitly known and have to be estimated and adapted. Nevertheless, even without the projection step, the effect of adaptivity is still clear to see, since only around the singularity, wavelets up to the maximum level are used.

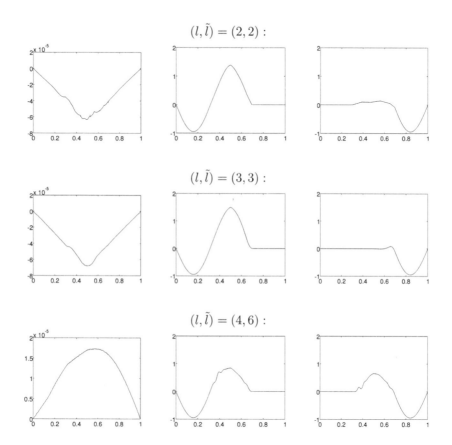

Figure 5.4: Pointwise error (left) and local approximations of the solution from the multiplicative Schwarz algorithm on Ω_0 (center) and Ω_1 (right).

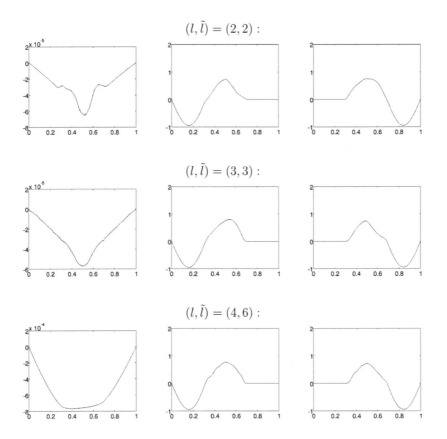

Figure 5.5: Pointwise error (left) and local approximations of the solution from the additive Schwarz algorithm on Ω_0 (center) and Ω_1 (right).

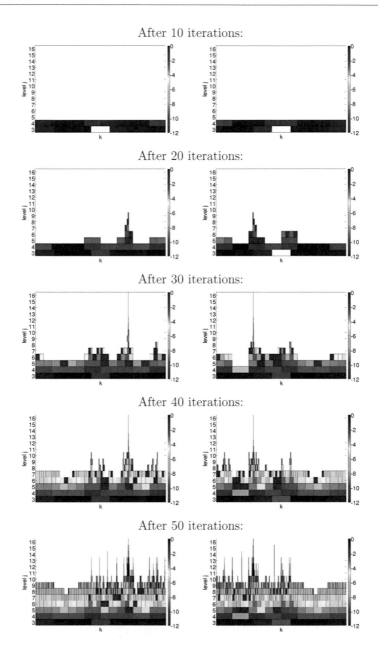

Figure 5.6: Active coefficients on Ω_0 (left) and Ω_1 (right) during the course of the additive Schwarz iteration with $(l, \tilde{l}) = (3, 3)$. The color indicates the magnitude of the coefficients in a logarithmic scale with basis 10.

5.3 Numerical tests on the L-shaped domain

So far, we have mostly confirmed the theoretically predicted behavior of the algorithms for one-dimensional test problems. Since partial differential equations in one space dimension are essentially ordinary differential equations, realistic applications of our methods, however, will take place in at least two space dimensions. Therefore, we now test the algorithms on the standard example of a two-dimensional polygonal domain, the L-shaped domain, which, in this section, will be denoted by Ω. We again make use of the reference problem

$$-\Delta u + u^3 = f \text{ in } \Omega, \quad u = 0 \text{ on } \partial\Omega$$

in weak form. At least for the standard Poisson equation without the nonlinear term u^3, it is known that the solution can we written as the sum of a regular part u_R and a singular part u_S, see the discussion in Section 3.5.1. Hence, on such polygonal domains, the solution has singularities that are imposed by the domain, even if the right-hand side is smooth. Adaptive methods have the potential to resolve such singularities better than uniform methods. Thus, as in [76, 82, 123] we choose as a solution to the model problem the singularity function

$$u(\theta, r) := u_S(\theta, r) := \eta(r) r^{2/3} \sin(\frac{2}{3}\theta),$$

see Figure 5.7. This function is written in polar coordinates (θ, r) around the origin, where η is a smooth truncation function so that $\eta(0) = 1$ and $\eta(x) = 0$ for $x \geq 1$. The right-hand side is calculated accordingly.

Recall from Subsection 3.5.1 that u_s is contained in $H^\vartheta(\Omega)$ only for $\vartheta < \frac{5}{3}$, but it belongs to all the Besov spaces $B_{\tau,\tau}^{2s+3/2}(\Omega)$, $\frac{1}{\tau} = s + \frac{1}{2}$. For the latter result, see the proof of [30, Th. 2.4]. This means that, in view of Proposition 3.2.1, we expect uniform methods to achieve at most a rate of $s = \frac{1}{3}$. The rate of nonlinear approximation, and therefore the benchmark for our adaptive algorithms, however, is only limited by the approximation power of the underlying wavelet basis or frame, see, e.g., Theorem 3.2.5. Therefore, the benchmark for our algorithms is the maximum rate of $s^* = \frac{l-t}{d}$. Although the expected rate increases with the order of the wavelets, we have already seen in the one-dimensional tests that the involved constants might get worse when applying wavelets of higher order. Therefore, from a quantitative point of view, the results achieved with piecewise quadratic wavelets, i.e., with parameters $(l, \tilde{l}) = (3, 3)$, were at least as good as the results we obtained when applying piecewise cubic wavelets and significantly better than the results with piecewise linear wavelets. This impression has been confirmed by other tests in two space dimensions, compare [75, Section 5.3]. Hence, in the following, we restrict ourselves to experiments with the piecewise quadratic wavelets with minimum level $j_0 = 3$ and maximum level $j_{\max} := 7$. The parameters in the algorithm were chosen as in the one-dimensional test example except that the relaxation parameter α in *LocSolve* was reduced to $\alpha := 0.1$

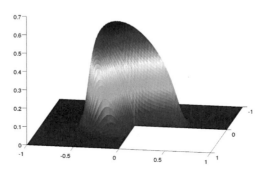

Figure 5.7: Exact solution $u(\theta, r)$ of the test problem on the L-shaped domain.

to assure convergence of the local Richardson iteration. Moreover, it has turned out empirically, that $c := 0.8$ is a good choice for the two-dimensional problem.

On the left-hand side in Figure 5.8, we compare the support sizes of the iterates to the approximate residual. We see that, after an initial phase, the additive Schwarz algorithm converges with approximately the expected rate. The multiplicative version performs a little better and, in some parts of the iteration, shows a slightly higher rate than expected. Regarding the computational time, which is shown on the right-hand side in Figure 5.8, we have to see that the rate cannot be observed as clearly as in the one-dimensional test examples. The algorithm runs slower than expected at the beginning of the iteration, but seems to catch up over the course of time. This might be an effect due to precomputations and storing the indices of intersecting wavelets, which leads to more time being spent early in the iteration, but gives a benefit in later stages.

In Figures 5.9 and 5.10, we see the local approximations of the solutions produced by the multiplicative and the additive Schwarz algorithm. In both cases, there seem to be no oscillations in the overlapping region, which indicates an economic representation of the solution.

The pointwise error from both the multiplicative and the additive Schwarz method is depicted in Figure 5.11. We see that, in both variants, the error is small except at the reentrant corner, where it peaks and takes a value around $5 \cdot 10^{-3}$. To explain this, it helps to take a look at Figures 5.12 and 5.13. There, we see the cubes belonging to set of active wavelet coefficients on the subdomains, occurring in the multiplicative Schwarz algorithm after 30 and 50 iterations, respectively. The additive Schwarz method shows a similar behavior, hence we omit the corresponding graphs. In the

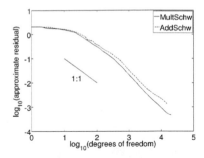 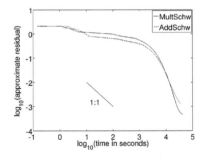

Figure 5.8: Degrees of freedom vs. ℓ_2-norm of the approximate residual (left) and computational time vs. ℓ_2-norm of the approximate residual (right) for the test on the L-shaped domain.

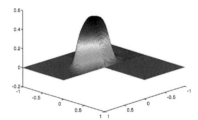 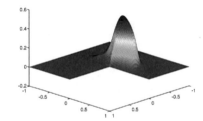

Figure 5.9: Local approximations of the solution from the multiplicative Schwarz algorithm on Ω_0 (left) and Ω_1 (right).

snapshot after 30 iterations, it can be observed that most degrees of freedom are spent on the relevant parts of the solution and that the singularity around the origin is also very well detected. After 50 iterations, it turns out that the active wavelet coefficients in this part the solution are filled up up to the maximum level $j_{\max} = 7$. This might also explain the behavior of the algorithm. To resolve the singularity even further, it would be necessary to increase $j_{\max}$ to a higher level. This however, causes additional administrative effort due to the way the building blocks are implemented, and thus slows down the algorithm significantly.

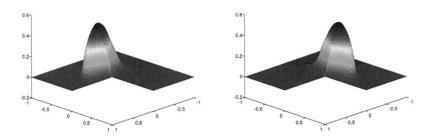

Figure 5.10: Local approximations of the solution from the additive Schwarz algorithm on Ω_0 (left) and Ω_1 (right).

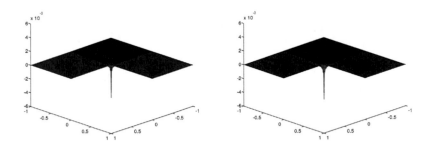

Figure 5.11: Pointwise error after 50 iterations in the multiplicative Schwarz method (left) and the additive Schwarz method (right).

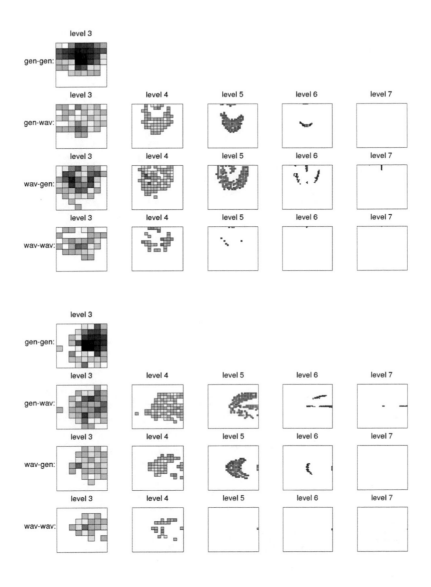

Figure 5.12: Active wavelet indices on subdomain Ω_0 (above) and Ω_1 (below) after 30 iterations of the multiplicative Schwarz method.

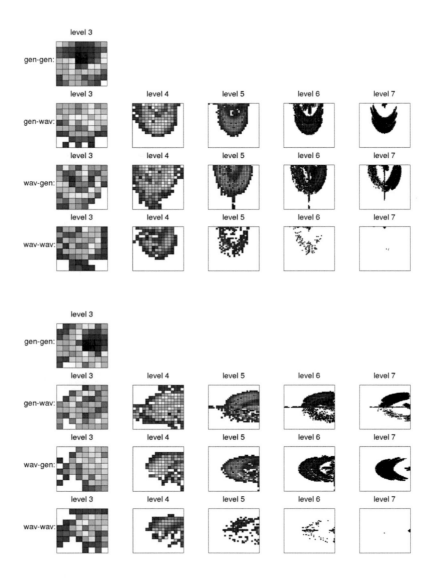

Figure 5.13: Active wavelet indices on subdomain Ω_0 (above) and Ω_1 (below) after 50 iterations of the multiplicative Schwarz method.

5.4 Some modifications and improvements

In the previous sections, we have seen that, at large, the adaptive methods realize the theoretically predicted optimal rate. Nevertheless, the computational performance is not yet fully satisfactory. The by far largest share of the computational time in the algorithms is needed for the evaluation of the nonlinearity, and therein particularly for the support prediction step, compare Subsection 4.2.3. Unfortunately, this step cannot be parallelized in a straightforward fashion, compare also the discussion in [75]. Therefore, the parallelization of the unmodified additive Schwarz method only gives a small benefit. Hence, following lines from loc. cit., a possible practical improvement is to call the support prediction step only after a fixed number of iterations.

We test this modification on the additive Schwarz method applied to the problem on the L-shaped domain from the previous section and with the same parameters otherwise. In Figure 5.14, we compare the modified version, where the support prediction step is only called in every fifth iteration, to the original additive Schwarz method. On the left-hand side, we draw the approximate residual compared to the degrees of freedom. In this benchmark, the modified method shows a behavior hardly distinguishable from the original method. This is good news, because it means that we do not significantly lose accuracy. On the right-hand side, we compare the computational time to the approximate residual. There, we observe that the modified version, denoted in the graph by *AddSchwMod*, is significantly faster by a factor of about 3. The modified method can further be accelerated by parallelizing the local solvers and the recover step with *OpenMP*, which is represented in the graph as *AddSchwPar*. In the example, this leads to a gain of around 17% of the total runtime. A much higher speed-up cannot be expected, because even in the modified version, roughly half of the computational effort is still spent on the support prediction that only runs on one node. However, from a practical point of view, the adjustments discussed in this section already lead to a significant speed-up.

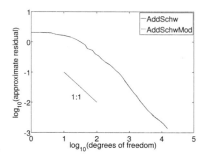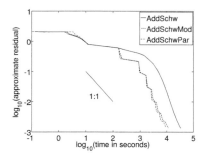

Figure 5.14: Degrees of freedom vs. ℓ_2-norm of the approximate residual (left) and computational time vs. ℓ_2-norm of the approximate residual (right) for the original and the modified additive Schwarz method, tested on the L-shaped domain.

Chapter 6

Adaptive Wavelet Schwarz methods for the Navier-Stokes equation

In the previous chapters, we have analyzed and tested Schwarz methods for semilinear elliptic equations. In this chapter, we outline how these techniques can be applied to another very important kind of equations from fluid dynamics, the stationary Navier-Stokes equation. Recall from Section 1.4, that this equation is given by

$$-\frac{1}{\text{Re}}\Delta u + (u \cdot \nabla)u + \nabla p = f, \quad \text{div}\, u = 0 \text{ in } \Omega, \quad u_{|\partial\Omega} = 0.$$

and that it describes the velocity u and the pressure p of a fluid in a domain $\Omega \subset \mathbb{R}^d$ under a given, constant external force f and a Reynolds number Re. In this chapter, we will work exclusively with its LeRay weak form

$$a(u,v) + \text{Re} \int_\Omega v \cdot (u \cdot \nabla)u \, dx = \text{Re} \int_\Omega f \cdot v \, dx, \quad v \in V,$$

which, for $d \leq 4$, was introduced in (1.25). Our approach applies to the case of small Reynolds numbers, which corresponds to laminar, e.g., smooth, non-turbulent flow, and incompressible fluids such as liquids. For an overview of the background and numerical treatment of the Navier-Stokes equation, we also refer to [114].

Wavelet methods for the numerical treatment of the Navier-Stokes equations have previously been studied in a range of papers. For instance, in [101], Petrov-Galerkin and collocation methods are compared and analyzed, albeit without convergence rates. A different approach can be taken when using divergence-free wavelets, because they allow us to use a more convenient weak formulation, in which the pressure term drops out, compare Section 1.4 and Subsection 2.3.2. In [50, 119], the application of such divergence-free wavelets in adaptive wavelet algorithms was studied. Moreover, since the LeRay weak form of the Stokes equation leads to an elliptic problem on the space of divergence-free wavelets, known iteration schemes for such problems can be adapted so that we have convergent and optimal algorithms, see [37, 74]. For related work on saddle point problems, we also refer to [31, 36]. However, as already in the case of the semilinear equations, the construction of the divergence-free wavelet bases on general domains, even on polygonal domains, poses a serious difficulty, for the same reasons as outlined in Subsection 2.2.4. Analogously to Subsection 2.2.5, these problems can often be circumvented by using a wavelet frame constructed with the help of an overlapping domain decomposition. We have seen in the previous chapters that this approach can in a natural way be combined with Schwarz methods. Hence, we describe the construction of an adaptive wavelet Schwarz method for the stationary, incompressible Navier-Stokes equation in two space dimensions, that includes the Stokes equation as a special case. With the help of techniques from [86], we show that the method is convergent, at least for sufficiently small Reynolds numbers. In addition, we prove that the method is asymptotically optimal with respect to the degrees of freedom.

The results and methods presented in this chapter, have, in similar but shorter form, already been published in the collaborative paper [38].

6.1 The basic principles of the algorithm

In this section, we outline the basic principles of a Schwarz method for the solution of the stationary Navier-Stokes equation. In the subsequent sections, we will construct an adaptive wavelet method based on these principles. The general approach of the method is similar to the methods from Chapter 4. The equation is linearized in each iteration by evaluating the nonlinearity $v \mapsto (v \cdot \nabla)v$ in the old iterate. Then, the linearized equation is solved approximately with an additive Schwarz method. In an idealized form, similarly to [86], the algorithm can be defined as follows, with the notation from Section 1.4 and $V_i := \{v \in (H_0^1(\Omega_i))^d,\ \mathrm{div}\, v = 0\}$ with $d \leq 4$.

Algorithm 15 AddSchw_NS1

$u^{(0)} := 0$
for $n = 0, 1, \ldots$ **do**
    for $i = 0, \ldots, m - 1$ **do**
        Compute $e_i^{(n)} \in V_i$ such that
$$a(e_i^{(n)}, v) = -a(u^{(n)}, v) - \mathrm{Re} \int_{\Omega_i} v \cdot (u^{(n)} \cdot \nabla) u^{(n)} + \mathrm{Re} \int_{\Omega_i} f \cdot v, \quad v \in V_i.$$
    end for
    $u^{(n+1)} := u^{(n)} + \omega \sum_{i=0}^{m-1} e_i^{(n)}$
end for

To formulate a convergence result for this algorithm, we need a few preparations. To begin with, we assume that the conditions from Theorem 1.4.2 are fulfilled, so that there exists a unique solution $u \in V$. Moreover, we fix a constant $M > 0$ such that $M \geq \|u\|$, where $\|\cdot\| := a(\cdot, \cdot)^{1/2}$ is the energy norm induced by the scalar product (1.24). Moreover, we denote by P_i the $a(\cdot, \cdot)$-orthogonal projector onto V_i.

Analogously to Proposition 4.1.7, it holds that

$$\theta := \|I - \omega \sum_{i=0}^{m-1} P_i\| < 1 \tag{6.1}$$

for all $\omega \in (0, \frac{2}{m})$. With these preparations at hand, we derive a first convergence result for the idealized algorithm, following lines from [86, Th. 4].

Proposition 6.1.1. *Let $d \leq 4$, C_{NS} be the constant from Theorem 1.4.2 and define $\rho := \theta + 3\omega m \operatorname{Re} C_{NS} M < 1$. Then, Algorithm 15 is convergent with*

$$\|u^{(n)} - u\| \leq \rho^n \|u^{(0)} - u\|.$$

Proof. The assertion is shown by an induction over n. The case $n = 0$ is obvious. Assume the induction hypothesis is valid for some $n \geq 0$. Let $b(\cdot, \cdot, \cdot)$ be the trilinear form from (1.26). Analogously to (4.17) and by using (1.25), it holds that for all $v \in V_i$, the local correction term $e_i^{(n)} \in V_i$ fulfills

$$\begin{aligned}
a(e_i^{(n)}, v) &= -a(u^{(n)}, v) - \operatorname{Re} b(v, u^{(n)}, u^{(n)}) + \operatorname{Re} f(v) \\
&= -a(u^{(n)}, v) - \operatorname{Re}(b(v, u^{(n)}, u^{(n)}) - b(v, u, u)) + \operatorname{Re} f(v) - b(v, u, u) \\
&= -a(u^{(n)}, v) - \operatorname{Re}(b(v, u^{(n)}, u^{(n)}) - b(v, u, u)) + a(u, v) \\
&= a(P_i(u - u^{(n)}), v) - \operatorname{Re}(b(v, u^{(n)}, u^{(n)}) - b(v, u, u))
\end{aligned}$$

From this, we obtain

$$e_i^{(n)} = P_i(u - u^{(n)}) - \operatorname{Re} \mathcal{L}_i^{-1}(b(\cdot, u^{(n)}, u^{(n)}) - b(\cdot, u, u)),$$

where $\mathcal{L}_i : V_i \to V_i'$, $\mathcal{L}_i(v)(w) = a(v, w)$. Because $b(\cdot, \cdot, \cdot)$ is linear in each of its components, this can be rewritten as

$$e_i^{(n)} = P_i(u - u^{(n)}) - \operatorname{Re} \mathcal{L}_i^{-1}(b(\cdot, u^{(n)}, u^{(n)} - u) - b(\cdot, u - u^{(n)}, u)).$$

Hence, by definition of $u^{(n+1)}$, it follows that

$$\begin{aligned}
&u - u^{(n+1)} \\
=&u - u^{(n)} - \omega \sum_{i=0}^{m-1} [P_i(u - u^{(n)}) - \operatorname{Re} \mathcal{L}_i^{-1}(b(\cdot, u^{(n)}, u^{(n)} - u) - b(\cdot, u - u^{(n)}, u))] \\
=&(I - \omega \sum_{i=0}^{m-1} P_i)(u - u^{(n)}) + \omega \operatorname{Re} \sum_{i=0}^{m-1} \mathcal{L}_i^{-1}(b(\cdot, u^{(n)}, u^{(n)} - u) - b(\cdot, u - u^{(n)}, u)).
\end{aligned}$$

Using the arguments from Lemma 1.4.1, with $d \leq 4$ and Ω replaced by Ω_i, we know that $\|\mathcal{L}_i^{-1} b(\cdot, v, w)\| \leq C_{NS} \|v\| \cdot \|w\|$. Combining this with (6.1) and $M \geq \|u\|$, we obtain

$$\|u^{(n+1)} - u\| \leq \theta \|u^{(n)} - u\| + \omega m \operatorname{Re} C_{NS}(\|u^{(n)}\| + \|u\|)\|u^{(n)} - u\|$$
$$\leq \theta \|u^{(n)} - u\| + \omega m \operatorname{Re} C_{NS}(\|u^{(n)} - u\| + 2\|u\|)\|u^{(n)} - u\|$$
$$\leq (\theta + \omega m \operatorname{Re} C_{NS}(\|u^{(n)} - u\| + 2M))\|u^{(n)} - u\|.$$

By the induction hypothesis, it holds that $\|u^{(n)} - u\| \leq \rho^n \|u^{(0)} - u\|$. In particular, we also have $\|u^{(n)} - u\| \leq \|u^{(0)} - u\| = \|u\| \leq M$. Thus, we conclude

$$\|u^{(n+1)} - u\| \leq (\theta + 3\omega m \operatorname{Re} C_{NS} M)\rho^n \|u^{(0)} - u\| \tag{6.2}$$
$$= \rho^{n+1} \|u^{(0)} - u\|.$$

$\square$

In the following, we outline how an adaptive wavelet method, based on this idealized algorithm, can be realized. For the construction of this algorithm, some modifications of the strategies applied in Chapter 4 are necessary, as we will explain in the next section.

6.2 An alternative projection strategy

To theoretically verify that the Schwarz methods from Chapter 4 are asymptotically optimal, we have applied a strategy to control the subdivision of the approximate solution into its local parts on the subdomains. This approach helps to avoid oscillations in the overlapping region that might cause the support sizes of the iterates, and thus the computational effort, to grow uncontrollably. Our method of choice for this task was the projector $\mathbf{P}$ from Subsection 4.3.1. Recall that the basic idea for the application of this projector to a vector $\mathbf{v} \in \ell_2(\Lambda)$ was to decompose $v = \mathbf{v}^\top \Psi^{H_0^t(\Omega)}$ with the help of a partition of unity $\{\sigma_i\}_{i=0}^{m-1}$ into $v = \sum_{i=0}^{m-1} \sigma_i v_i$ and to calculate $\mathbf{w}_i$ as the expansion coefficients of $\sigma_i v$ in the local Riesz bases $\Psi^{H_0^t(\Omega_i)}$ for $H_0^t(\Omega_i)$. Then, $\mathbf{P}\mathbf{v}$, simply defined as $(\mathbf{w}_0, \ldots, \mathbf{w}_{m-1})$, was another representation of $v = \mathbf{v}^\top \Psi^{H_0^t(\Omega)} = (\mathbf{P}\mathbf{v})^\top \Psi^{H_0^t(\Omega)}$, but with explicitly controlled local parts.

This approach, however, cannot directly be applied when working with divergence-free wavelets. This is because for $v \in V$, the pointwise multiplication of all its components with σ_i is no longer divergence-free in general. Hence, it cannot be represented in the wavelet Riesz basis $\Psi^{(i)}$ for V_i. Therefore, we outline in this section an alternative construction of such a projector and its properties following [38, Section 3.1]. Roughly speaking, the basic idea of this strategy, explained for the case of $m = 2$ subdomains Ω_0, Ω_1, is to let the parts of a function that are supported in the overlapping region be represented almost only by wavelets from the basis on Ω_0, except for a small strip around the inner boundaries $\partial\Omega_0 \cap \Omega$. Only the remainder is represented in the Riesz basis from Ω_1. It seems comprehensible that, by such a strategy, we can avoid oscillations in the overlapping region that might endanger the sparsity of the representation of a function in the wavelet frame. This approach is compatible with the use of divergence-free wavelets, but could be applied for other wavelet constructions as well.

For the definition of this alternative projection strategy, we need to make an additional assumption on the nature of the domain decomposition. To formulate this condition, for $\delta > 0$ and $0 \le i \le m-1$, we define

$$\Omega_i(-\delta) := \{x \in \Omega_i,\ B(x,\delta) \cap \Omega \subset \Omega_i\},$$

where $B(x; \delta)$ is the closed ball of radius δ around x. This set can be understood as the subdomain Ω_i with a strip of width δ being removed at the inner interfaces with other subdomains, but not on the outer interfaces. With this definition, we assume that we can find such a $\delta > 0$, so that, for all $\lambda \in \Lambda_i$ and $0 \le i \le m-1$, we have

$$\operatorname{diam\ supp} \psi_\lambda^{(i)} \cup \operatorname{supp} \tilde{\psi}_\lambda^{(i)} \le \delta/2, \tag{6.3}$$

where $\Psi^{(i)} = \{\psi_\lambda^{(i)}\}_{\lambda \in \Lambda_i}$ are the local divergence-free Riesz bases for V_i from Section 2.2.6 with dual bases $\tilde{\Psi}^{(i)} = \{\tilde{\psi}_\lambda^{(i)}\}_{\lambda \in \Lambda_i}$, and, simultaneously,

$$\Omega \subset \bigcup_{i=0}^{m-1} \Omega_i(-(1 + \frac{m-1}{2})\delta). \tag{6.4}$$

Remark 6.2.1. *The condition (6.4) can be interpreted in such a way that the overlapping region of the subdomains is sufficiently large compared to the supports of the wavelets. The latter size, and thereby the $\delta > 0$ for which the condition has to be guaranteed, can be reduced by choosing a higher minimal level of the original wavelet bases. However, for some standard decompositions of polygonal domains such as the decomposition of the L-shaped domain from Example 2.2.15, this property (6.4) does not hold for any $\delta > 0$, compare also the discussion in [123, Remark 6.2].*

Under the assumptions formulated above, we can show the following lemma, that serves as a preparation for the construction of a suitable projection method.

Lemma 6.2.2. *Let $\Lambda_i^\delta := \{\lambda \in \Lambda_i,\ \operatorname{supp} \psi_\lambda^{(i)} \cap \Omega_i(-\delta) \ne \emptyset\}$. Then, the mapping $v \mapsto (\langle \tilde{\psi}_\lambda^{(i)}, v \rangle_{V' \times V})_{\lambda \in \Lambda_i^\delta}$ is a bounded linear operator from V to $\ell_2(\Lambda_i^\delta)$. Moreover, for each $v \in V$, $v - \sum_{\lambda \in \Lambda_i^\delta} \langle \tilde{\psi}_\lambda^{(i)}, v \rangle \psi_\lambda^{(i)}$ vanishes on $\Omega_i(-\delta)$.*

Proof. Let $\sigma : \Omega \to \mathbb{R}$ be a smooth function that is equal to one on $\Omega_i(-\frac{1}{3}\delta)$ and equal to zero on $\Omega \setminus \Omega_i(-\frac{1}{6}\delta)$. Such a function can be constructed due to Assumption (6.4) with the help of a smooth cut-off function as in Section 2.2.5, compare also [123, Section 6.1.5]. In particular, $\{\sigma, (1 - \sigma)\}$ is a partition of unity with respect to the domain decomposition $\Omega = \Omega_i \cup (\Omega \setminus \Omega_i(-\delta/2))$. By Lemma 2.2.10, Corollary 2.2.12 and Lemma 2.2.18, each $v \in V$ can therefore be decomposed as $v = v_i + w_i$ with $v_i \in V_i$, $w_i \in \{w \in H_0^1(\Omega \setminus \Omega_i(-\delta/2))^d\}$ and $\|v\|_{H^1(\Omega)^d} \approx \|v_i\|_{H^1(\Omega)^d} + \|w_i\|_{H^1(\Omega)^d}$. By Assumption (6.3), $(\langle \tilde{\psi}_\lambda^{(i)}, v \rangle_{V' \times V})_{\lambda \in \Lambda_i^\delta}$ only depends on $v_{|\Omega_i(-\delta/2)}$, and we have $v = v_i$ on $\Omega_i(-\delta/2)$. Because $\tilde{\Psi}^{(i)}$ is a Riesz basis, and thus a frame, the corresponding analysis operator F is bounded, hence it holds that

$$\|(\langle \tilde{\psi}_\lambda^{(i)}, v \rangle_{V' \times V})_{\lambda \in \Lambda_i^\delta}\|_{\ell_2(\Lambda_i^\delta)} = \|(\langle \tilde{\psi}_\lambda^{(i)}, v_i \rangle_{V' \times V})_{\lambda \in \Lambda_i^\delta}\|_{\ell_2(\Lambda_i^\delta)} \lesssim \|v_i\|_{H^1(\Omega)^d} \lesssim \|v\|_{H^1(\Omega)^d},$$

which shows the first assertion. Because $\tilde{\Psi}^{(i)}$ is the dual of $\Psi^{(i)}$, by Theorem 2.1.2 and the above, it holds that $v_i = \sum_{\lambda \in \Lambda_i} \langle \tilde{\psi}_\lambda^{(i)}, v_i \rangle \psi_\lambda^{(i)}$, hence, by choice of Λ_i^δ and $v = v_i$ on $\Omega(-\delta/2)$, it follows that $v = \sum_{\lambda \in \Lambda_i^\delta} \langle \tilde{\psi}_\lambda^{(i)}, v \rangle \psi_\lambda^{(i)}$ on $\Omega_i(-\delta)$, which completes the proof. $\qquad \square$

With the help of this result, we can successively define the following mappings, that generalize the basic idea described in the introduction of this section to the case of $m \geq 2$ subdomains. To do so, we set $Z_{-1} = H_{-1} = 0$ and, iteratively, for $0 \leq i \leq m - 1$, we define

$$H_i : \ell_2(\Lambda_0) \times \ldots \times \ell_2(\Lambda_i) \to V, \quad (\mathbf{v}_0, \ldots, \mathbf{v}_i) \mapsto \sum_{j=0}^{i} \mathbf{v}_j^\top \Psi^{(j)},$$

and

$$Z_i : V \to \ell_2(\Lambda_0) \times \ldots \times \ell_2(\Lambda_i), v \mapsto (Z_{i-1}v, \langle \tilde{\psi}_\lambda^{(i)}, v - H_{i-1}Z_{i-1}v \rangle)_{\lambda \in \Lambda_i^\delta}).$$

Finally, we set $Z := Z_{m-1}$, considered as a mapping from V to $\ell_2(\Lambda)$ and define $\tilde{\mathbf{P}} := ZF^*$, where $F^* : \ell_2(\Lambda) \to V$ is the synthesis operator corresponding to the frame $\Psi = \bigcup_{i=0}^{m-1} \Psi^{(i)}$ for V, compare Subsection 2.1.2. To see that this operator is suitable, at least in theory, for the application as a projection step within an adaptive wavelet algorithm similarly to Subsection 4.3.1, we have to verify that $\tilde{\mathbf{P}}$ is indeed a bounded projector on $\ell_2(\Lambda)$ that only changes the representation of a function $v \in V$ in the wavelet frame Ψ, but not the function itself. These properties are summarized in the following proposition.

Proposition 6.2.3. *The mappings Z_i are bounded operators. For $v \in V$, $v - H_i Z_i v$ vanishes on $\bigcup_{j=0}^{i} \Omega_j(-(1 + \frac{i}{2})\delta)$. In particular, $\tilde{\mathbf{P}}$ is a bounded projector on $\ell_2(\Lambda)$ and it holds that $F^* \tilde{\mathbf{P}} = F^*$, i.e., $(\tilde{\mathbf{P}}\mathbf{v})^\top \Psi = \mathbf{v}^\top \Psi$ for all $\mathbf{v} \in \ell_2(\Lambda)$.*

Proof. Because the $\Psi^{(j)}$ are Riesz bases, the mappings H_i are bounded. By induction over i, we see that even the Z_i are bounded. For $i = -1$, this holds by definition. To show the induction step, it suffices to note that by the induction hypothesis and Lemma 6.2.2, the mapping

$$v \mapsto \langle (\tilde{\psi}_\lambda^{(i)}, v - H_{i-1}Z_{i-1}v) \rangle_{\lambda \in \Lambda_i^\delta}$$

is bounded from V to $\ell_2(\Lambda_i^\delta)$. In particular, Z is bounded as well, and thus $\tilde{\mathbf{P}}$ as a composition of bounded operators. To see that $v - H_i Z_i$ vanishes on $\bigcup_{j=0}^i \Omega_j(-(1 + \frac{i}{2})\delta)$, we also use induction over i, where the case $i = -1$ holds by definition of Z_{-1} and H_{-1}. To show the inductive step, note that by definition of H_i and Z_i, we have the equation

$$v - H_i Z_i v = (v - H_{i-1} Z_{i-1} v) - \sum_{\lambda \in \Lambda_i^\delta} \langle \tilde{\psi}_\lambda^{(i)}, v - H_{i-1} Z_{i-1} v \rangle \psi_\lambda^{(i)}.$$

By the induction hypothesis, $v - H_{i-1} Z_{i-1}$ vanishes on $\bigcup_{j=0}^i \Omega_j(-(1 + \frac{i-1}{2})\delta)$. Combining the induction hypothesis with Assumption (6.3) gives that the second term vanishes on

$$\bigcup_{j=0}^i \Omega_j(-(1 + \frac{i-1}{2})\delta + \frac{1}{2}\delta) = \bigcup_{j=0}^i \Omega_j(-(1 + \frac{i}{2})\delta),$$

hence even $v - H_i Z_i v$ vanishes on $\bigcup_{j=0}^i \Omega_j(-(1 + \frac{i}{2})\delta)$. In particular, because $H_{m-1} = F^*$ and $Z_{m-1} = Z$, it follows that $v = F^* Z v$ on $\bigcup_{j=0}^{m-1} \Omega_j(-(1+\frac{m-1}{2})\delta)$. With Assumption (6.4), we conclude that $F^* Z = I$ and therefore $F^* \tilde{\mathbf{P}} = F^* Z F^* = F^*$. Furthermore, we have $\tilde{\mathbf{P}}^2 = Z F^* Z F^* = Z F^* = \tilde{\mathbf{P}}$. $\square$

One main purpose of the projector $\mathbf{P}$ in Chapter 4 was to ensure that, if the solution $u \in H_0^t(\Omega)$ has a representation $u = \mathbf{u}^\top \Psi^{H_0^s(\Omega)}$ with $\mathbf{u} \in \mathcal{A}_{\mathcal{AT}}^s$, the iterates $\mathbf{u}^{(n)}$ produced by the algorithm remain uniformly bounded in $\mathcal{A}_{\mathcal{AT}}^s$, so it holds that $\|\mathbf{u}^{(n)}\|_{\mathcal{A}_{\mathcal{AT}}^s} \lesssim \|\mathbf{u}\|_{\mathcal{A}_{\mathcal{AT}}^s}$. To show this, it was necessary to assume that $\mathbf{P}$ is bounded on $\mathcal{A}_{\mathcal{AT}}^s$. To proceed in a similar fashion, we have to make an analogous assumption on $\tilde{\mathbf{P}}$, i.e., from now on, let $\tilde{\mathbf{P}}$ be bounded on $\mathcal{A}_{\mathcal{AT}}^s$ for a sufficiently large $s > 0$. In the following, we explain how it can be verified, given additional assumptions essentially on the underlying basis.

In a first step, we observe that the problem can be reduced to showing the boundedness of some smaller matrices on a tree approximation space.

Lemma 6.2.4. *Let, for $0 \le i, j \le m - 1$, the matrices*

$$\mathbf{B}^{(i,j)} := (\langle \tilde{\psi}_\lambda^{(i)}, \psi_\mu^{(j)} \rangle)_{\lambda \in \Lambda_i^\delta, \mu \in \Lambda_j}$$

be bounded from $\mathcal{A}_{\mathcal{T}_j}^s$ to $\mathcal{A}_{\mathcal{T}_i}^s$. Then, $\tilde{\mathbf{P}}$ is bounded on $\mathcal{A}_{\mathcal{AT}}^s$.

Proof. We show by an induction over i that $Z_i F^*$ is bounded on $\mathcal{A}_{\mathcal{T}_0}^s \times \dots \mathcal{A}_{\mathcal{T}_i}^s$. This shows the assertion, because $\tilde{\mathbf{P}} = Z_{m-1} F^*$ and $\mathcal{A}_{\mathcal{AT}}^s = \mathcal{A}_{\mathcal{T}_0}^s \times \dots \times \mathcal{A}_{\mathcal{T}_{m-1}}^s$. For $i = 0$, we have

$$Z_0 F^* \mathbf{v} = (\langle \tilde{\psi}_\lambda^{(0)}, v \rangle)_{\lambda \in \Lambda_0^\delta} = \sum_{j=0}^{m-1} \mathbf{B}^{(0,j)} \mathbf{v},$$

from which the assertion follows by the boundedness of the $\mathbf{B}^{(0,j)}$. For the inductive step, note that

$$Z_i F^* \mathbf{v} = (Z_{i-1} F^* \mathbf{v}, \sum_{j=0}^{m-1} \mathbf{B}^{(i,j)} \mathbf{v} - \sum_{j=0}^{i-1} \mathbf{B}^{(i,j)}(Z_{i-1} F^* \mathbf{v})),$$

hence the assertion follows by the induction hypothesis and the boundedness of the $\mathbf{B}^{(i,j)}$. $\qquad\square$

In view of the above lemma, we now fix some $0 \leq i, j \leq m - 1$ and abbreviate $\mathbf{B} := \mathbf{B}^{(i,j)}$. To show that this matrix is bounded on the tree approximation spaces, we need the following assumptions.

Assumption 6.2.5. *Let $\mathbf{B}$ as defined above, $\varepsilon > 0$ and $\mathbf{v}_\varepsilon \in \ell_2(\Lambda_i)$. We state the following assumptions:*

(i) For some $s^* > s$, $\mathbf{B}$ is s^*-compressible.

(ii) Let $\lambda \in \Lambda_i^\delta$, $\mu \in \Lambda_j$, $l \in \mathbb{Z}$ with $|\lambda| \geq l \geq |\mu|$ and $\operatorname{supp} \tilde{\psi}_\lambda^{(i)} \cap \operatorname{supp} \psi_\mu^{(j)} \neq \emptyset$. Then, there exits a $\vartheta \in \Lambda_i^\delta$ with $|\vartheta| = l$ and $\operatorname{supp} \tilde{\psi}_\vartheta^{(i)} \cap (\operatorname{supp} \tilde{\psi}_\lambda^{(i)} \cap \operatorname{supp} \psi_\mu^{(j)}) \neq \emptyset$.

(iii) Let $\mathbf{w}_\varepsilon := \mathbf{APPLY}[\mathbf{B}, \mathbf{v}_\varepsilon, \varepsilon/2] \in \ell_2(\Lambda_i)$. If $\mu \in \operatorname{supp} \mathbf{v}_\varepsilon$, $\lambda \in \operatorname{supp} \mathbf{w}_\varepsilon$ and $\vartheta \in \Lambda_i^\delta$ with $|\mu| \leq |\vartheta| \leq |\lambda|$ and $\operatorname{supp} \tilde{\psi}_\vartheta^{(i)} \cap \operatorname{supp} \psi_\mu^{(j)} \neq \emptyset$, then $\vartheta \in \operatorname{supp} \mathbf{w}_\varepsilon$.

Under these assumptions, that will be discussed in a little more detail below, we can show that $\mathbf{P}$ is bounded on $\mathcal{A}_{\mathcal{AT}}^s$.

Proposition 6.2.6. *Let Assumption 6.2.5 be fulfilled. Then, $\tilde{\mathbf{P}}$ is bounded on $\mathcal{A}_{\mathcal{AT}}^s$.*

Proof. First of all, with the arguments from Lemma 6.2.2, together with $\tilde{\Psi}^{(i)}$ and $\Psi^{(j)}$ being Riesz bases, we observe that $\mathbf{B}$ defines a bounded operator from $\ell_2(\Lambda_j)$ to $\ell_2(\Lambda_i^\delta) \subset \ell_2(\Lambda_i)$. To discuss the boundedness of $\mathbf{B}$ on the tree approximation spaces, let $\mathbf{v} \in \mathcal{A}_{\mathcal{T}_j}^s$ be given. For any $\varepsilon > 0$, we have to find a vector $\mathbf{w}_\varepsilon \in \ell_2(\Lambda_i)$ with $\|\mathbf{B}\mathbf{v} - \mathbf{w}_\varepsilon\|_{\ell_2(\Lambda_i)} \leq \varepsilon$ and $\#\mathcal{T}(\operatorname{supp} \mathbf{w}_\varepsilon) \lesssim \varepsilon^{-1/s}\|\mathbf{v}\|_{\mathcal{A}_{\mathcal{T}_j}^s}^{1/s}$, where $\mathcal{T}(\operatorname{supp} \mathbf{w}_\varepsilon)$, as in Subsection 4.2.1, is the smallest tree in Λ_i containing $\mathbf{w}_\varepsilon$. In a first step, we define

$$\mathbf{v}_\varepsilon := \mathbf{T\_COARSE}[\varepsilon/(2\|\mathbf{B}\|), \mathbf{v}]$$

with the coarsening method from Subsection 4.2.1 and $\|\mathbf{B}\| < \infty$ being the operator norm on the ℓ_2-spaces. Because $\mathbf{T\_COARSE}$ coincides with $\mathbf{AT\_COARSE}$ in the case $m = 1$, from Lemma 4.2.3, we have

$$\|\mathbf{B}\| \cdot \|\mathbf{v} - \mathbf{v}_\varepsilon\|_{\ell_2(\Lambda_j)} \leq \varepsilon/2, \quad \#\operatorname{supp} \mathbf{v}_\varepsilon \lesssim \varepsilon^{-1/s}\|\mathbf{v}\|_{\mathcal{A}_{\mathcal{T}_j}^s}^{1/s} \quad \|\mathbf{v}_\varepsilon\|_{\mathcal{A}_{\mathcal{T}_j}^s} \lesssim \|\mathbf{v}\|_{\mathcal{A}_{\mathcal{T}_j}^s}, \quad (6.5)$$

where, in the last two estimates, $\|\mathbf{B}\|$ is hidden in the implicit constants. Given part (i) of Assumption 6.2.5, we can make use of the approximate matrix-vector multiplication from Subsection 4.2.2 and calculate

$$\mathbf{w}_\varepsilon := \mathbf{APPLY}[\mathbf{B}, \mathbf{v}_\varepsilon, \varepsilon/2] \in \ell_2(\Lambda_i).$$

By Proposition 4.2.6, we have $\|\mathbf{B}\mathbf{v}_\varepsilon - \mathbf{w}_\varepsilon\|_{\ell_2(\Lambda_i)} \leq \varepsilon/2$, thus with (6.5), it follows that $\|\mathbf{B}\mathbf{v} - \mathbf{w}_\varepsilon\|_{\ell_2(\Lambda_i)} \leq \varepsilon$. In addition, with $|\cdot|_{\ell^w_\tau(\Lambda_i)} \approx \|\cdot\|_{\mathcal{A}^s} \lesssim \|\cdot\|_{\mathcal{A}^s_{\mathcal{T}_j}}$, (4.19) and (6.5), we have

$$\# \operatorname{supp} \mathbf{w}_\varepsilon \lesssim \varepsilon^{-1/s} \|\mathbf{v}_\varepsilon\|^{1/s}_{\mathcal{A}^s_{\mathcal{T}_j}} \lesssim \varepsilon^{-1/s} \|\mathbf{v}\|^{1/s}_{\mathcal{A}^s_{\mathcal{T}_j}}. \tag{6.6}$$

To show $\#\mathcal{T}(\operatorname{supp} \mathbf{w}_\varepsilon) \lesssim \varepsilon^{-1/s} \|\mathbf{v}\|^{1/s}_{\mathcal{A}^s_{\mathcal{T}_j}}$, we consider the set

$$\tilde{\mathcal{T}}_C := \{\theta \in \Lambda_i, \exists \lambda \in \operatorname{supp} \mathbf{w}_\varepsilon : |\lambda| \geq |\theta|, \operatorname{dist}(\operatorname{supp} \psi^{(i)}_\lambda, \operatorname{supp} \psi^{(i)}_\theta) \leq C 2^{-|\theta|}\}.$$

Because the divergence-free wavelets are local, compare Subsection 2.2.6, it holds that $\operatorname{diam} \operatorname{supp} \psi^{(i)}_\theta \lesssim 2^{-|\theta|}$. Hence, for a sufficiently large fixed parameter C, $\tilde{\mathcal{T}}_C$ contains $\mathcal{T}(\operatorname{supp} \mathbf{w}_\varepsilon)$. Thus, we have to find an upper bound for $\#\tilde{\mathcal{T}}_C$. The basic strategy we will follow is to show that, for any $\theta \in \tilde{\mathcal{T}}_C$, we find a $\psi^{(j)}_\gamma$, $\gamma \in \operatorname{supp} \mathbf{v}_\varepsilon$ or $\tilde{\psi}^{(i)}_\vartheta$ with $\vartheta \in \operatorname{supp} \mathbf{w}_\varepsilon$ that are on the same level as θ and lie close to ψ_θ. Then, using the locality of the primal and dual wavelets, we conclude that the cardinality of $\tilde{\mathcal{T}}_C$ is of the same magnitude as $\# \operatorname{supp} \mathbf{v}_\varepsilon + \# \operatorname{supp} \mathbf{w}_\varepsilon$.

For $\lambda \in \operatorname{supp} \mathbf{w}_\varepsilon$, there exists a $\mu \in \operatorname{supp} \mathbf{v}_\varepsilon$ with $\operatorname{supp} \tilde{\psi}^{(i)}_\lambda \cap \operatorname{supp} \psi^{(j)}_\mu \neq \emptyset$. This is because otherwise, in the sum $(\mathbf{B}\mathbf{v}_\varepsilon)_\lambda = \sum_\theta \langle \tilde{\psi}^{(i)}_\lambda, \psi^{(j)}_\theta \rangle (\mathbf{v}_\varepsilon)_\theta$, all the terms vanish. Then, the index λ does not play a role in the approximation of $\mathbf{B}\mathbf{v}_\varepsilon$.

Let $\theta \in \tilde{\mathcal{T}}_C$ be given and $\lambda \in \operatorname{supp} \mathbf{w}_\varepsilon$ as in the definition of $\tilde{\mathcal{T}}_C$. Then, as above, let $\mu \in \operatorname{supp} \mathbf{v}_\varepsilon$ with $\operatorname{supp} \tilde{\psi}^{(i)}_\lambda \cap \operatorname{supp} \psi^{(j)}_\mu \neq \emptyset$. Consider first the case $|\theta| \leq |\mu|$. Because $\operatorname{supp} \mathbf{v}_\varepsilon$ is tree-structured, we can choose a predecessor $\gamma \in \operatorname{supp} \mathbf{v}_\varepsilon$ of μ, or $\gamma = \mu$, so that $|\gamma| = |\theta|$. Because γ is a predecessor of μ, it holds that

$$\operatorname{dist}(\operatorname{supp} \psi^{(j)}_\mu, \operatorname{supp} \psi^{(j)}_\gamma) \lesssim 2^{-|\theta|}$$

Because $\operatorname{supp} \psi^{(j)}_\mu$ intersects $\operatorname{supp} \tilde{\psi}^{(i)}_\lambda$, which, in turn, intersects $\operatorname{supp} \psi^{(i)}_\lambda$ and the diameter of the supports of these wavelets are all bounded by a constant times $2^{-|\theta|}$, it follows that $\operatorname{dist}(\operatorname{supp} \psi^{(i)}_\lambda, \operatorname{supp} \psi^{(j)}_\gamma) \lesssim 2^{-|\theta|}$. Then, by definition $\tilde{\mathcal{T}}_C$ and $|\theta| = |\gamma|$, it holds that

$$\operatorname{dist}(\operatorname{supp} \psi^{(i)}_\theta, \operatorname{supp} \psi^{(j)}_\gamma) \lesssim 2^{-|\theta|}. \tag{6.7}$$

In the other case, $|\theta| > |\mu|$, because $\operatorname{supp} \tilde{\psi}^{(i)}_\lambda \cap \operatorname{supp} \psi^{(j)}_\mu \neq \emptyset$, by part (ii) in Assumption 6.2.5, we can find a $\vartheta \in \Lambda^\delta_i$ with $|\theta| = |\vartheta|$ and

$$\operatorname{supp} \tilde{\psi}^{(i)}_\vartheta \cap (\operatorname{supp} \tilde{\psi}^{(i)}_\lambda \cap \operatorname{supp} \psi^{(j)}_\mu) \neq \emptyset.$$

Then, by part (iii) of Assumption 6.2.5, it holds that $\vartheta \in \operatorname{supp} \mathbf{w}_\varepsilon$ and, moreover,

$$\operatorname{dist}(\operatorname{supp} \psi_\theta^{(i)}, \operatorname{supp} \tilde{\psi}_\vartheta^{(i)}) \lesssim 2^{-|\theta|}. \tag{6.8}$$

Because the divergence-free wavelets are local, for each $\gamma \in \operatorname{supp} \mathbf{v}_\varepsilon$, $|\gamma| = |\theta|$, there is only a uniformly bounded number of $\theta \in \tilde{\mathcal{T}}_C$, for which (6.7) can hold. Similarly for each $\vartheta \in \operatorname{supp} \mathbf{w}_\varepsilon$, there is only a uniformly bounded number of candidates for $\theta \in \tilde{\mathcal{T}}_C$, $|\theta| = |\vartheta|$, so that (6.8) is valid. Hence, in abstract terms, we have defined a mapping $\tilde{\mathcal{T}}_C \to \operatorname{supp} \mathbf{v}_\varepsilon \cup \operatorname{supp} \mathbf{w}_\varepsilon$ so that the cardinality of the preimage of each element in $\operatorname{supp} \mathbf{v}_\varepsilon \cup \operatorname{supp} \mathbf{w}_\varepsilon$ is uniformly bounded. Therefore, we finally have

$$\#\tilde{\mathcal{T}}_C \lesssim \# \operatorname{supp} \mathbf{v}_\varepsilon + \# \operatorname{supp} \mathbf{w}_\varepsilon \lesssim \varepsilon^{-1/s} \|\mathbf{v}\|_{\mathcal{A}_{T_j}^s}^{1/s},$$

which is what we need to show to conclude the boundedness of $\mathbf{B}$ on the tree approximation spaces. □

Remark 6.2.7. *Let us make a few remarks on Assumption 6.2.5. If the construction of the divergence-free wavelets in Subsection 2.2.6 starts with spline wavelets, then even the resulting divergence-free wavelets are compactly supported splines, since they are derived from the original basis on the interval by means of integration and differentiation. Hence, we expect that with the approach from [104, Section 4], we can show that $\mathbf{B}$ is s^*-compressible for some $s^* > \frac{l-t}{d}$. In view of the results from Proposition 3.4.7, this s^* is larger than the parameter s, for which $\mathbf{u} \in \mathcal{A}_{AT}^s$ can be expected to hold. The point (ii) is essentially an assumption on the wavelets. If the dual bases have nested support, ϑ can be chosen so that $\operatorname{supp} \tilde{\psi}_\vartheta^{(i)}$ contains $\operatorname{supp} \tilde{\psi}_\lambda^{(i)}$. Regarding the third part of the assumption, note that the sparse matrices $\mathbf{B}_j$ approximating $\mathbf{B}$ within the routine **APPLY** from Subsection 4.2.2 are typically constructed using estimates for the off-diagonal decay of the entries of $\mathbf{B}$. This means that these matrices $\mathbf{B}_j$ are constructed by cutting away the parts of the matrix $\mathbf{B}$ that correspond to a larger difference of the levels, compare also [104, Section 4.5]. Therefore, we can expect that if $\vartheta \in \Lambda_i^\delta$ with an intermediate level $|\mu| \le |\vartheta| \le |\lambda|$ and $\operatorname{supp} \tilde{\psi}_\vartheta^{(i)} \cap \operatorname{supp} \psi_\mu^{(j)} \ne \emptyset$, then $\vartheta \in \operatorname{supp} \mathbf{w}_\varepsilon$. Even if, by chance, $(\mathbf{w}_\varepsilon)_\vartheta = 0$, this index will be included in the derivation of the upper bound (6.6) for the support of $\mathbf{w}_\varepsilon$, which is why assuming $\vartheta \in \operatorname{supp} \mathbf{w}_\varepsilon$ really is not a restriction.*

In the formulation of the adaptive wavelet method in the subsequent section, for simplicity, we will write the inexact application of the projector $\tilde{\mathbf{P}}$ to a vector in $\ell_2(\Lambda)$ as the computation of an inexact matrix-vector with the method **AT_APPLY** from Subsection 4.2.2. However, as described in Subsection 4.3.1, in practice we would like to avoid the explicit use of the dual Riesz bases, i.e., of the wavelets $\tilde{\psi}_\lambda^{(i)}$, because they typically do not possess a spline structure. Instead, we could take a similar approach as with the projector $\mathbf{P}$, because the application of the operators Z_i basically amounts to calculating the expansion coefficients of $v - H_{i-1} Z_{i-1}$, which can, similarly to Subsection 4.3.1, be written as the iterative solution of linear systems.

Alternatively, since this projection step is only needed for the theoretical verification of optimality, and in view of the numerical results from Chapter 5, leaving away the implementation of $\tilde{\mathbf{P}}$ completely might practically be even more advantageous.

6.3 The adaptive wavelet method

In this section, we formulate and analyze an adaptive wavelet method based on the idealized algorithm from Section 6.1. Since the construction of isotropic divergence-free wavelets with Dirichlet boundary conditions that we apply is limited to the two-dimensional case, we restrict ourselves to $d = 2$. Similarly to the adaptive algorithms from Chapter 4, it will be necessary to switch between the norm of a function in V and the $\ell_2(\Lambda)$-norm of its coefficients in a wavelet frame $\Psi := \Psi^{V(\Omega)}$. As in Subsection 4.3.2, because the energy norm $\|\!|\cdot|\!\|$ is equivalent to the Sobolev norm on $(H_0^t(\Omega))^2$ and because the synthesis operator F^* is bounded, there exists a constant $\tilde{K}_1 > 0$ so that

$$\|\!|\mathbf{v}^\top \Psi|\!\| = \|\!|F^*\mathbf{v}|\!\| \leq \tilde{K}_1 \|\mathbf{v}\|_{\ell_2(\Lambda)} \tag{6.9}$$

holds uniformly in $\mathbf{v} \in \ell_2(\Lambda)$. Moreover, because $\tilde{\mathbf{P}} = ZF^*$ and Z is bounded, there exists a constant $\tilde{K}_2 > 0$ such that

$$\|\tilde{\mathbf{P}}\mathbf{v}\|_{\ell_2(\Lambda)} \leq \tilde{K}_2 \|\!|F^*\mathbf{v}|\!\| = \|\!|\mathbf{v}^\top \Psi|\!\|. \tag{6.10}$$

Furthermore, with ρ defined as in Proposition 6.1.1, we set $\tilde{\rho} := \frac{1}{2}(1 + \rho)$. With these preparations at hand, the methods **AT_COARSE** and **AT_APPLY** from Section 4.2, and the constant C^* from Lemma 4.2.3, we are ready to define the adaptive wavelet version of Algorithm 15.

Algorithm 16 AddSchw_NS2[ε]

% Let $l^* \in \mathbb{N}$ be minimal with $\tilde{\rho}^{l^*} \leq \frac{1}{2K_1 K_2} \frac{1}{2C^*+1} \tilde{\rho}$.

% Let $\varepsilon_n := \tilde{\rho}^n M$, $n \in \mathbb{N}$.

$u^{(0)} := 0$

$n := 0$

while $\varepsilon_n \geq \varepsilon$ **do**

  $v^{(n,0)} := u^{(n)}$

  for $l = 0, \ldots, l^* - 1$ **do**

    for $i = 0, \ldots, m-1$ **do**

      Compute $\tilde{e}_i^{(n,l)} \in V_i$ as an approximation to $e_i^{(n,l)} \in V_i$ from

$$a(e_i^{(n,l)}, v) = -a(v^{(n,l)}, v) - \mathrm{Re} \int_{\Omega_i} v \cdot (v^{(n,l)} \cdot \nabla) v^{(n,l)} + \mathrm{Re} \int_{\Omega_i} f \cdot v, \ v \in V_i$$

      with tolerance $\|e_i^{(n,l)} - \tilde{e}_i^{(n,l)}\| \leq \frac{1-\rho}{2m\omega} \varepsilon_n \tilde{\rho}^l$.

    end for

    $v^{(n,l+1)} := v^{(n,l)} + \omega \sum_{i=0}^{m-1} \tilde{e}_i^{(n,l)}$

  end for

  $\tilde{\mathbf{u}}^{(n+1)} := \mathbf{AT\_APPLY}[\tilde{\mathbf{P}}, \mathbf{v}^{(n,l^*)}, \frac{1}{2K_1} \frac{1}{2C^*+1} \varepsilon_{n+1}]$

  $\mathbf{u}^{(n+1)} := \mathbf{AT\_COARSE}[\tilde{\mathbf{u}}^{(n+1)}, \frac{1}{K_1} \frac{2C^*}{2C^*+1} \varepsilon_{n+1}]$

  $n := n + 1$

end while

$\mathbf{u}_\varepsilon := u^{(n)}$

$u_\varepsilon := \mathbf{u}_\varepsilon^\top \Psi$

In this formulation, we have not yet explained how the subproblems can be solved in practice. This will be discussed in the subsequent section. Assuming, for the time being, that we have at hand such a local solver, we can show the following convergence result. The optimality of the algorithm will be discussed afterwards.

Theorem 6.3.1. *Let $\varepsilon > 0$, the Reynolds number $\mathrm{Re} > 0$ be sufficiently small so that $\rho < 1$ and let the Assumptions (6.3) and (6.4) be fulfilled. Then, for the iterates $u^{(n)}$ from Algorithm 16, it holds that*

$$\|u^{(n)} - u\| \leq \varepsilon_n, \tag{6.11}$$

and, in particular, $\|u_\varepsilon - u\| \leq \varepsilon$.

Proof. We proceed by induction over n, similarly to the proof of Proposition 4.3.3. By choice of M and $u^{(0)}$, the assertion is true in the case $n = 0$. To proceed the induction, assume that (6.11) holds for some $n \geq 0$. We show by an inner induction over l that $\|v^{(n,l)} - u\| \leq \tilde{\rho}^l \varepsilon_n$. Because of $v^{(n,0)} = u^{(n)}$, the case $l = 0$ holds by the outer induction hypothesis (6.11). To proceed from l to $l + 1$, note that

$$v^{(n,l+1)} - u = \left(v^{(n,l)} + \omega \sum_{i=0}^{m-1} d_i^{(n,l)} - u \right) + \omega \sum_{i=0}^{m-1} (\tilde{d}_i^{(n,l)} - d_i^{(n,l)}).$$

The first term is the result of one exact step of the exact Schwarz iteration as in Algorithm 15. Thus, as in (6.2), the error in this step is reduced by the factor ρ. Combining this with the error bound for the solution of the local subproblems and using the inner induction hypothesis, $\|v^{(n,l)} - u\| \leq \tilde{\rho}^l \varepsilon_n$, we obtain

$$\|v^{(n,l+1)} - u\| \leq \rho\|v^{(n,l)} - u\| + \omega m \frac{1-\rho}{2m\omega}\varepsilon_n\tilde{\rho}^l \leq \rho\tilde{\rho}^l\varepsilon_n + \frac{1-\rho}{2}\varepsilon_n\tilde{\rho}^l = \tilde{\rho}^{l+1}\varepsilon_n,$$

completing the inner induction. In particular, by choice of l^* and the estimate (6.10), it follows that

$$\|\tilde{\mathbf{P}}v^{(n,l^*)} - \tilde{\mathbf{P}}u\|_{\ell_2(\Lambda)} \leq \tilde{K}_2\|v^{(n,l^*)} - u\| \leq \tilde{K}_2\tilde{\rho}^{l^*}\varepsilon_n \leq \frac{1}{2\tilde{K}_1}\frac{1}{2C^*+1}\tilde{\rho}\varepsilon_n.$$

By the tolerances for the call **AT_APPLY**, together with $\tilde{\rho}\varepsilon_n = \varepsilon_{n+1}$, this gives

$$\begin{aligned}
\|\tilde{\mathbf{u}}^{(n+1)} - \tilde{\mathbf{P}}u\|_{\ell_2(\Lambda)} &\leq \|\tilde{\mathbf{u}}^{(n+1)} - \tilde{\mathbf{P}}v^{(n,l^*)}\|_{\ell_2(\Lambda)} + \|\tilde{\mathbf{P}}v^{(n,l^*)} - \tilde{\mathbf{P}}u\|_{\ell_2(\Lambda)} \\
&\leq \frac{1}{2\tilde{K}_1}\frac{1}{2C^*+1}\varepsilon_{n+1} + \frac{1}{2\tilde{K}_1}\frac{1}{2C^*+1}\varepsilon_{n+1} \\
&= \frac{1}{\tilde{K}_1}\frac{1}{2C^*+1}\varepsilon_{n+1}.
\end{aligned} \tag{6.12}$$

Combining this estimate with the tolerance for **AT_COARSE**, we have

$$\begin{aligned}
\|\mathbf{u}^{(n+1)} - \tilde{\mathbf{P}}u\|_{\ell_2(\Lambda)} &\leq \|\mathbf{u}^{(n+1)} - \tilde{\mathbf{u}}^{(n+1)}\|_{\ell_2(\Lambda)} + \|\tilde{\mathbf{u}}^{(n+1)} - \tilde{\mathbf{P}}u\|_{\ell_2(\Lambda)} \\
&\leq \frac{1}{\tilde{K}_1}\frac{2C^*}{2C^*+1}\varepsilon_{n+1} + \frac{1}{\tilde{K}_1}\frac{1}{2C^*+1}\varepsilon_{n+1} \\
&= \frac{1}{\tilde{K}_1}\varepsilon_{n+1}.
\end{aligned}$$

Hence, using (6.9) and $(\tilde{\mathbf{P}}u)^\top \Psi = u$ from Proposition 6.2.3, it follows that

$$\|u^{(n+1)} - u\| \leq \tilde{K}_1\|\mathbf{u}^{(n+1)} - \tilde{\mathbf{P}}u\|_{\ell_2(\Lambda)} \leq \varepsilon_{n+1},$$

which completes the proof. $\qquad \square$

Without specifying how exactly the subproblems are solved, we can already show that the algorithm is asymptotically optimal with respect to the degrees of freedom. To do so, we proceed as in Lemma 4.3.7, using the properties of the coarsening method.

Proposition 6.3.2. *Let $u = \mathbf{u}^\top \Psi$ with $\mathbf{u} \in \mathcal{A}_{AT}^s$ and $\tilde{\mathbf{P}}$ be bounded on $\mathcal{A}_{AT}^s$. Then, all iterates $\mathbf{u}^{(n)}$ from Algorithm 16 are contained in $\mathcal{A}_{AT}^s$ with*

$$\|\mathbf{u}^{(n)}\|_{\mathcal{A}_{AT}^s} \lesssim \|\mathbf{u}\|_{\mathcal{A}_{AT}^s}, \quad \#\operatorname{supp}\mathbf{u}^{(n)} \lesssim \varepsilon_n^{-1/s}\|\mathbf{u}\|_{\mathcal{A}_{AT}^s}^{1/s}.$$

Proof. Combining Lemma 4.2.3 with the bound (6.12) gives $\|\mathbf{u}^{(n+1)}\|_{\mathcal{A}_{\mathcal{AT}}^s} \lesssim \|\tilde{\mathbf{P}}\mathbf{u}\|_{\mathcal{A}_{\mathcal{AT}}^s}$ and $\#\operatorname{supp}\mathbf{u}^{(n+1)} \lesssim \varepsilon_{n+1}^{-1/s}\|\tilde{\mathbf{P}}\mathbf{u}\|_{\mathcal{A}_{\mathcal{AT}}^s}^{1/s}$, from which the assertion follows from the boundedness of $\tilde{\mathbf{P}}$ on $\mathcal{A}_{\mathcal{AT}}^s$. □

Although we do not strictly prove that the overall algorithm is optimal with respect to the computational effort as well, a discussion of this aspect is part of the next subsection, where we outline how to solve the subproblems.

6.4 On the numerical solution of the subproblems

In this section, we discuss the solution of the local subproblems arising in Algorithm 16. In discretized form, these local subproblems can be written as

$$\mathbf{A}^{(i,i)}\mathbf{e}_i^{(n,l)} = -(\mathbf{A}\mathbf{v}^{(n,l)})_{|\Lambda_i} - \operatorname{Re}\mathbf{G}(\mathbf{v}^{(n,l)})_{|\Lambda_i} + \operatorname{Re}\mathbf{f}_{|\Lambda_i} =: \mathbf{r}, \qquad (6.13)$$

where $\mathbf{A}^{(i,j)} = \{a(\psi_\lambda, \psi_\mu)\}_{\lambda\in\Lambda_i, \mu\in\Lambda_j}$ is the (i,j)-th diagonal block of the stiffness matrix $\mathbf{A}$, $\mathbf{G}: \ell_2(\Lambda) \to \ell_2(\Lambda)$ and $\mathbf{f} \in \ell_2(\Lambda)$ are the discrete nonlinearity and the right-hand side, respectively, as introduced in Subsection 2.3.2. In principle, the subproblems can be treated in a similar fashion as the subproblems arising in the Schwarz algorithms for semilinear equations, see Subsections 4.3.3 and 4.4.2. This is because the matrix $\mathbf{A}^{(i,i)}$ is positive definite on $\ell_2(\Lambda_i)$, see Subsection 2.3.2, and the right-hand side is a vector in $\ell_2(\Lambda_i)$. Thus, this problem is basically in the same form as the problems (4.42) and (4.59). Hence, we restrict ourselves to outlining the main principles and possible differences compared to the approach from Chapter 4. The basic idea, also used in [111] for linear problems, is to apply the inexact Richardson iteration from [26] for the solution of the local subproblems. Given a sufficiently small relaxation parameter $\alpha > 0$ and an arbitrary initial guess $\mathbf{y}^{(0)} \in \ell_2(\Lambda_i)$, the exact Richardson iteration is given by

$$\mathbf{y}^{(r+1)} = \mathbf{y}^{(r)} + \alpha(\mathbf{r} - \mathbf{A}^{(i,i)}\mathbf{y}^{(r)}).$$

This iteration is convergent to the exact solution $\mathbf{e}_i^{(n,l)}$ of the problem (6.13), even if the evaluation of $\mathbf{A}^{(i,i)}\mathbf{y}^{(r)}$ and the calculation of the right-hand side $\mathbf{r}$ can only be performed inexactly. For the matrix-vector products with $\mathbf{A}$ and its main diagonal blocks $\mathbf{A}^{(i,i)}$, we can use the method **AT_APPLY** from Subsection 4.2.2. The required compressibility of this matrix has been verified in [74]. For the approximation of $\mathbf{r}$, we furthermore have to calculate $\mathbf{f}$ and the nonlinearity $\mathbf{G}(\mathbf{v}^{(n,l)})$, at least up to any given tolerance. As before, a suitable method for the evaluation of the right-hand side is assumed to exist, compare Subsection 4.2.4. The nonlinearity takes a different, quadratic, form compared to the Nemitsky operators treated in Subsection 4.2.3. However, if the underlying system is a wavelet basis, the evaluation of such terms is covered by the theory in [27]. There, sparsity and complexity estimates as in Proposition 4.2.15 could be shown even for the nonlinearity from the Navier-Stokes equation.

For wavelet frames, such a result has not yet been strictly verified. However, we expect this to be possible. This is because for nonlinearities of one argument, we have seen in [75, 76] that the arguments from [27] carry over to the case of aggregated wavelet frames. Assuming such an evaluation routine, the arguments from Subsection 4.3.3 can be reiterated so that even the optimality proof from Theorem 4.3.10 regarding the computational effort can be adapted to the Navier-Stokes equation.

Chapter 7

Adaptive Newton-Schwarz method

In the course of this thesis, we have developed adaptive wavelet methods for nonlinear problems that all made use of a simple, but effective linearization strategy. Based on the idea of the Picard iteration and the methods in [85], we have generated linear subproblems by evaluating the nonlinearity in the old iterate and moving it to the right-hand side of the equation, compare Section 4.1. This can be considered a linerization of order zero in the sense that we approximate the nonlinearity in each iteration by a constant function. A natural further step to take is to ask what can be gained if we work with a linearization of the next higher order one, which means that we approximate the nonlinearity by a linear, non-constant function. This leads us to the area of Newton's method. Such methods are a well-known technique for the solution of nonlinear equations. A main difference of these schemes compared to the simpler linearization strategies as those we have applied in Chapter 4 is that an idealized Newton method is quadratically convergent, at least in a neighborhood of the solution. For a general overview of Newton's method and its applications, we also refer to [51, 77].

In the context of adaptive wavelet methods for partial differential equations, a method of this kind has first been analyzed theoretically in [27] and implemented in [121]. There, it was outlined how such a method can be realized so that it is convergent and asymptotically optimal given that the nonlinearity is stable and monotone. This method, however, relies on the existence of a suitable wavelet Riesz basis on the domain on which the problem is posed. As pointed out in Chapter 2, the construction of such a basis poses serious difficulties that can, in many cases, be avoided when instead working with a wavelet frame generated with the help of an overlapping domain decomposition. Moreover, we have seen in Chapter 4 that overlapping Schwarz methods are particularly well-suited to be combined with this construction. Hence, in the following, we will bring together adaptive wavelet Newton-type methods with Schwarz domain decomposition techniques. This will lead us to an adaptive wavelet Newton-Schwarz method. Although Newton-Schwarz methods are well-known in the context of general nonlinear equations, see, e.g., [3, 4, 13], to the best of our knowledge, the application of this particular kind of methods in the context of adaptive wavelet solvers for nonlinear partial differential equations has not yet been studied. Although some details of the analysis still remain open in this chapter, it may serve as a starting point for further research, in particular because Newton's method might be applicable to a larger range of nonlinearities than those considered in this thesis.

7.1 The basic approach

To describe the Newton-Schwarz method, we first introduce the basic principles of Newton's method in our setting, following lines from [27]. Thereafter, we discuss the combination with an additive Schwarz method.

Newton's method is typically formulated as an algorithm for finding the zeros of a multidimensional function. Hence, to make this approach compatible with our problem, we rewrite the weak form (4.1) of the semilinear problem as

$$\mathcal{R}(u) = 0, \tag{7.1}$$

where, for $v \in H_0^t(\Omega)$, $\mathcal{R}(v) := \mathcal{L}v + \mathcal{G}(v) - f \in H^{-t}(\Omega)$ denotes the residual. In the following, we assume that the nonlinearity $\mathcal{G}$ is a Nemitsky operator stemming from a monotone function $G : \mathbb{R} \to \mathbb{R}$.

Starting from an initial guess $\bar{u}^{(0)} \in H_0^t(\Omega)$ for the solution u to problem (7.1), the basic idea of Newton's method is to successively calculate a sequence of approximations by linearizing this equation in $\bar{u}^{(n)}$ and finding $\bar{u}^{(n+1)}$ as the solution of the linearized equation. With the help of the Fréchet derivative $D\mathcal{R}$ of $\mathcal{R}$, compare Equation (1.13), we therefore approximate $\mathcal{R}(\bar{u}^{(n+1)})$ by its first order Taylor approximation

$$\mathcal{R}(\bar{u}^{(n+1)}) \doteq \mathcal{R}(\bar{u}^{(n)}) + D\mathcal{R}(\bar{u}^{(n)})(\bar{u}^{(n+1)} - \bar{u}^{(n)}).$$

Assuming that $D\mathcal{R}(\bar{u}^{(n)})$ is invertible, we can define the next approximation $\bar{u}^{(n+1)}$ as the unique zero of this expression,

$$\bar{u}^{(n+1)} = \bar{u}^{(n)} - D\mathcal{R}(\bar{u}^{(n)})^{-1}\mathcal{R}(\bar{u}^{(n)}). \tag{7.2}$$

The following result shows that, under the monotonicity and a growth assumption on the nonlinearity and at least locally, this iteration is well-defined and convergent. This result is a consequence of Theorem 1.2.2 and [27, Section 8], which, in turn, makes use of the general convergence theory of Newton's method for partial differential equations from [51].

Proposition 7.1.1. *Assume that the function G is monotone and fulfills a growth condition (1.14) for some $n^* \geq 2$. Then, the problem (7.1) has a unique solution $u \in H_0^t(\Omega)$ and there exists a neighborhood $U = B_\delta(u)$ such that for each $v \in U$, the Fréchet derivative $D\mathcal{R}$ is an isomorphism from $H_0^t(\Omega)$ to $H^{-t}(\Omega)$. Moreover, for any initial approximation $\bar{u}^{(0)} \in U$, the iterates are quadratically convergent, i.e., $\|u^{(n)} - u\|_{H^t(\Omega)} \downarrow 0$ and there exists a constant $\beta > 0$ such that*

$$\|\bar{u}^{(n+1)} - u\|_{H^t(\Omega)} \leq \beta \|\bar{u}^{(n)} - u\|_{H^t(\Omega)}^2. \tag{7.3}$$

By rewriting the exact Newton iteration (7.2), while assuming that $\bar{u}^{(n)} \in U$, one step of this scheme amounts to solving for $\bar{v}^{(n)} := \bar{u}^{(n+1)} - \bar{u}^{(n)} \in H_0^t(\Omega)$ the linear problem

$$D\mathcal{R}(\bar{u}^{(n)})\bar{v}^{(n)} = -\mathcal{R}(\bar{u}^{(n)}). \tag{7.4}$$

The basic idea of the Newton-Schwarz algorithm is to approximate the solution of this linear problem with a Schwarz method. In our case, we will work with the additive Schwarz method because of its inherent parallelism. However, in principle, we could also use a multiplicative Schwarz solver. In any case, to control the overall error when applying the Schwarz solvers to the linear system, analogously to Chapter 4, it is necessary to make sure that the involved linear mapping induces a scalar product and an energy norm equivalent to the standard Sobolev norm. This is because the convergence of the Schwarz methods can be shown in this induced energy norm, whereas Newton's scheme is convergent in the Sobolev norm. To combine these two schemes and to show the convergence of this combined algorithm, we have to make sure these norms are compatible. Thus, for $v \in U$, we define

$$a_v(y, z) := \langle D\mathcal{R}(v)y, z \rangle, \quad y, z \in H_0^t(\Omega).$$

For the semilinear equations we work with, this expression can also be written as

$$a_v(y, z) = \langle \mathcal{L}y, z \rangle + \langle G'(v)y, z \rangle, \quad y, z \in H_0^t(\Omega),$$

see [27, Section 8]. In the following lemma, we establish some important properties of this bilinear form.

Lemma 7.1.2. *Let the assumptions from Proposition 7.1.1 be fulfilled. For each v in a neighborhood U of the solution u, $a_v(\cdot, \cdot)$ defines a scalar product on $H_0^t(\Omega)$. Moreover, the induced energy norm $\|w\|_v := (a_v(w, w))^{1/2}$ is equivalent to the Sobolev norm on $H_0^t(\Omega)$ with constants that can be chosen independently of $v \in U$. In addition, with $\Psi := \Psi^{H_0^t(\Omega)}$ being a wavelet frame for $H_0^t(\Omega)$ and with the projector $\mathbf{P}$ from Subsection 4.3.1, we have the norm equivalence*

$$c_E \|\mathbf{P}\mathbf{w}\|_{\ell_2(\Lambda)} \leq \|\mathbf{w}^\top \Psi\|_v \leq C_E \|\mathbf{P}\mathbf{w}\|_{\ell_2(\Lambda)}, \quad \mathbf{w} \in \ell_2(\Lambda) \tag{7.5}$$

with constants $c_E, C_E > 0$ independent from $v \in U$.

Proof. The proof partly follows lines from [27]. Because $D\mathcal{R} = D\mathcal{F}$ with $\mathcal{F}$ from Equation (1.16), from the estimate (1.17), it follows that

$$c\|w\|_{H^t(\Omega)} \leq \|D\mathcal{R}(v)(w)\|_{H^{-t}(\Omega)} \leq C_G(\|v\|_{H^t(\Omega)})\|w\|_{H^t(\Omega)}, \quad v \in U, w \in H_0^t(\Omega),$$

in a neighborhood $U = B_\delta(u)$. Because U is bounded, the upper bound can be made independent from $v \in U$. Hence, on the one hand it holds that

$$\|w\|_v^2 = a_v(w, w) = \langle D\mathcal{R}(v)w, w \rangle \leq \|D\mathcal{R}(v)w\|_{H^{-t}(\Omega)}\|w\|_{H^t(\Omega)} \approx \|w\|_{H^t(\Omega)}^2.$$

On the other hand, using that G is monotone and $\mathcal{L}$ is elliptic, we conclude

$$\|w\|_v^2 = \langle \mathcal{L}w, w \rangle + \langle G'(v)w, w \rangle \geq \langle \mathcal{L}w, w \rangle \approx \|w\|_{H^t(\Omega)}^2.$$

This shows the norm equivalence between the $\| \cdot \|_v$-norm and the Sobolev norm, uniformly in $v \in U$, and that $a_v(\cdot, \cdot)$ is positive definite and thus a scalar product. Moreover, combining this result with the estimate (4.34) shows the norm equivalence (7.5) and thereby completes the proof. $\qquad\square$

With the help of the above bilinear form, we can reformulate (7.4) as

$$a_{\bar{u}^{(n)}}(\bar{v}^{(n)}, w) = -\langle \mathcal{R}(\bar{u}^{(n)}), w \rangle, \quad w \in H_0^t(\Omega). \tag{7.6}$$

Analogously to Subsection 4.1.2, we apply an additive Schwarz iteration to this problem. Starting from an approximation $\bar{v}^{(n,k)}$ to $\bar{v}^{(n)}$, we calculate the local correction terms $e_i^{(n,k)} \in H_0^t(\Omega_i)$, $0 \leq i \leq m-1$ from the subproblems

$$a_{\bar{u}^{(n)}}(\bar{v}^{(n,k)} + e_i^{(n,k)}, w) = -\langle \mathcal{R}(\bar{u}^{(n)}), w \rangle, \quad w \in H_0^t(\Omega_i)$$

or, equivalently,

$$a_{\bar{u}^{(n)}}(e_i^{(n,k)}, w) = -a_{\bar{u}^{(n)}}(\bar{v}^{(n,k)}, w) - \langle \mathcal{R}(\bar{u}^{(n)}), w \rangle, \quad w \in H_0^t(\Omega_i).$$

With a relaxation parameter $\omega > 0$, the next global iterate can then be calculated as

$$\bar{v}^{(n,k+1)} := \bar{v}^{(n,k)} + \omega \sum_{i=0}^{m-1} e_i^{(n,k)}. \tag{7.7}$$

The convergence of this scheme for the problems (7.6) is guaranteed by the following lemma. Moreover, it shows that the error reduction factor in each step of this iteration is bounded away from one independent from $\bar{u}^{(n)} \in U$. This will greatly simplify the realization of the method, because it has the consequence that the number of inner iterations (7.7) needed to achieve a given error reduction can be chosen independently from the current global iterate $u^{(n)}$.

Lemma 7.1.3. *For the inner iterates $\bar{v}^{(n,k)}$ from Equation (7.7), it holds that*

$$\|\bar{v}^{(n,k+1)} - \bar{v}^{(n)}\|_{\bar{u}^{(n)}} \leq \rho_{\bar{u}^{(n)}} \|\bar{v}^{(n,k)} - \bar{v}^{(n)}\|_{\bar{u}^{(n)}},$$

where

$$\rho_{\bar{u}^{(n)}} := \|I - \omega(P_{0,\bar{u}^{(n)}} + \ldots + P_{m-1,\bar{u}^{(n)}})\|_{\bar{u}^{(n)}}.$$

and $P_{i,v}$ is the $a_v(\cdot, \cdot)$-orthogonal projector onto $H_0^t(\Omega_i)$. Moreover, if $\omega \in (0, \frac{2}{m})$, there exists a constant $\rho < 1$ such that $\rho_v \leq \rho$ for all $v \in U$.

Proof. The first assertion follows analogously to the proof of Proposition 4.1.6 from the relation $\bar{v}^{(n)} - \bar{v}^{(n,k+1)} = (I - \omega \sum_{i=0}^{m-1} P_{i,\bar{u}^{(n)}})(\bar{v}^{(n)} - \bar{v}^{(n,k)})$.

To show the second part, we define the operator $T_v := \sum_{i=0}^{m-1} P_{i,v}$, which is symmetric and positive definite with respect to the inner product $a_v(\cdot, \cdot)$. This can be verified as in the proof of Proposition 4.1.7. Hence, it follows that

$$\rho_v = \|I - \omega T_v\|_v = \max\{1 - \omega \lambda_{\min}(T_v), \omega \lambda_{\max}(T_v) - 1\}. \tag{7.8}$$

Since T_v is a sum of m orthogonal projectors, it is $\|T\|_v \leq m$. In particular, the maximum eigenvalue $\lambda_{\max}(T_v)$ is bounded from above by m. To find a lower bound for $\lambda_{\min}(T_v)$, we proceed following lines from [88, Ch. 8]. First of all, since the subdomains are assumed to be overlapping, each $w \in H_0^t(\Omega)$ can be decomposed as $w = \sum_{i=0}^{m-1} w_i$ with $w_i \in H_0^t(\Omega_i)$ and $\sum_{i=0}^{m-1} \|w_i\|_{H^t(\Omega)} \lesssim \|w\|_{H^t(\Omega)}$. Moreover, because $\|\cdot\|_v \approx \|\cdot\|_{H^t(\Omega)}$ holds uniformly in $v \in U$, see Lemma 7.1.2, there is a constant $\zeta \in \mathbb{R}$ such that $\sum_{i=0}^{m-1} \|w_i\|_v^2 \leq \zeta \|w\|_v^2$. A calculation analogous to the estimate (4.12) and the fact that the $P_{i,v}$ are $a_v(\cdot, \cdot)$-orthogonal projectors show that

$$\|w\|_v^2 \leq \left(\sum_{i=0}^{m-1} \|P_{i,v} w\|_v^2 \right)^{1/2} \left(\sum_{i=0}^{m-1} \|w_i\|_v^2 \right)^{1/2}$$

$$\leq \left(\sum_{i=0}^{m-1} a_v(P_{i,v} w, P_{i,v} w) \right)^{1/2} \zeta^{1/2} \|w\|_v$$

$$= a_v(T_v w, w)^{1/2} \zeta^{1/2} \|w\|_v.$$

Hence, it is $a_v(w, w) \leq \zeta \cdot a_v(T_v w, w)$. Using the Courant principle for the minimal eigenvalue, from this we obtain

$$\lambda_{\min}(T_v) = \inf_{w \in H_0^t(\Omega), w \neq 0} \frac{a_v(T_v w, w)}{a_v(w, w)} \geq \inf_{w \in H_0^t(\Omega), w \neq 0} \frac{a_v(T_v w, w)}{\zeta \cdot a_v(T_v w, w)} = \frac{1}{\zeta}.$$

Thus, together with (7.8), it follows that

$$\rho_v = \max\{1 - \omega \lambda_{\min}(T_v), \omega \lambda_{\max}(T_v) - 1\} \leq \max\left\{1 - \frac{\omega}{\zeta}, \omega m - 1\right\} =: \rho,$$

which is smaller than one for all $\omega \in (0, \frac{2}{m})$. $\qquad\square$

The next step will be to combine the above ideas in a consolidated Newton-Schwarz method, taking into account that all steps can only be performed inexactly. A challenge in the construction is to properly coordinate the tolerances for the different steps involved so that, altogether, we can ensure that the algorithm is convergent and asymptotically optimal, at least with respect to the degrees of freedom. This will be the focus of the next section.

7.2 The adaptive Newton-Schwarz algorithm

With the principles described in the previous section, we are now ready to define an adaptive wavelet version of the Newton-Schwarz method. Similarly to the Schwarz methods from Chapter 4, projection and coarsening strategies will be applied to assure an asymptotically optimal balance between the accuracy and the degrees of freedom. The numerical solution of the subproblems will be discussed in the subsequent section. In the formulation of the algorithm and in the analysis, we denote by $\Psi = \Psi^{H_0^t(\Omega)}$ a wavelet frame for $H_0^t(\Omega)$ constructed as a union of local wavelet Riesz bases $\Psi^{(i)} = \Psi^{H_0^t(\Omega_i)}$ for $H_0^t(\Omega_i)$, compare Section 2.2.5. Moreover, we will make use of the constants c_E and C_E from the norm equivalence (7.5). In addition, from the boundedness of the synthesis operator F^*, see Subsection 2.1.2, we infer the existence of a constant $M_1 := \|F^*\|_{\ell_2(\Lambda) \to H_0^t(\Omega)} < \infty$, for which the estimate

$$\|\mathbf{v}^\top \Psi\|_{H^t(\Omega)} \leq M_1 \|\mathbf{v}\|_{\ell_2(\Lambda)} \tag{7.9}$$

holds uniformly in $\mathbf{v} \in \ell_2(\Lambda)$, and from Lemma 4.3.1, we conclude that there exists a constant $M_2 > 0$ such that

$$\|\mathbf{P}\mathbf{v}\|_{\ell_2(\Lambda)} \leq M_2 \|\mathbf{v}^\top \Psi\|_{H^t(\Omega)}. \tag{7.10}$$

These constants will help us to switch between the different norms involved. With this notation fixed, the algorithm reads as follows.

Algorithm 17 NewtonSchwarz

% Choose tolerances $\varepsilon_n \searrow 0$ such that $\delta > \varepsilon_0 \geq \|u - u^{(0)}\|_{H^t(\Omega)}$,
% where $U = B_\delta(u)$ and $\varepsilon_{n+1} > (2C^* + 1)M_1 M_2 \beta \varepsilon_n^2$.
% Let $\tilde{\rho} := \frac{1}{2}(1 + \rho)$.
$n := 0$
while $\varepsilon_n \leq \varepsilon$ **do**
  $v^{(n,0)} = 0$
  % Choose $k^* \in \mathbb{N}$ minimal such that $\tilde{\rho}^{k^*} \leq \frac{1}{4\varepsilon_n} \frac{c_E}{C_E} \frac{1}{M_1 M_2} \left(\frac{1}{2C^*+1} \varepsilon_{n+1} - M_1 M_2 \beta \varepsilon_n^2\right)$
  for $k = 0, \ldots, k^* - 1$ **do**
    for $i = 0, \ldots, m - 1$ **do**
      Compute $\tilde{e}_i^{(n,k)} = (\tilde{\mathbf{e}}_i^{(n,k)})^\top \Psi^{(i)} \in H_0^t(\Omega_i)$ as approximation to $e_i^{(n,k)} \in H_0^t(\Omega_i)$
      from
$$a_{u^{(n)}}(e_i^{(n,k)}, w) = -a_{u^{(n)}}(v^{(n,k)}, w) - \mathcal{R}(u^{(n)})(w), \quad w \in H_0^t(\Omega_i)$$
      up to tolerance $\|e_i^{(n,k)} - \tilde{e}_i^{(n,k)}\|_{u^{(n)}} \leq \frac{1-\rho}{m\omega} C_E M_2 \varepsilon_n \tilde{\rho}^k$.
    end for
    $\mathbf{v}^{(n,k+1)} = \mathbf{v}^{(n,k)} + \omega \sum_{i=0}^{m-1} \tilde{\mathbf{e}}_i^{(n,k)}$
  end for
  $\hat{\mathbf{u}}^{(n+1)} := \mathbf{AT\_APPLY}[\mathbf{P}, \mathbf{u}^{(n)} + \mathbf{v}^{(n,k^*)}, \frac{1}{2M_1}\left(\frac{1}{2C^*+1}\varepsilon_{n+1} - M_1 M_2 \beta \varepsilon_n^2\right)]$
  $\mathbf{u}^{(n+1)} := \mathbf{AT\_COARSE}[\hat{\mathbf{u}}^{(n+1)}, \frac{1}{M_1}\frac{2C^*}{2C^*+1}\varepsilon_{n+1}]$
end while

Let us shortly discuss some aspects of this method. The choice of the tolerances ε_n within the algorithm is not completely determined. The condition on these tolerances allows us to choose up to quadratically decreasing values ε_n. In this case, however, also the tolerances for the local solver decrease quadratically. This is expected to increase the effort for the solution of the subproblems, so it is not clear in advance which choice of the sequence $(\varepsilon_n)_{n \geq 0}$ within the allowed bound actually is best from a computational point of view, compare also the discussion in [121, Ch. 8]. This aspect will be discussed in a little more detail in Section 7.4. However, for any choice of tolerances satisfying the condition stated within the algorithm, we can show the convergence of the method. The aspect of optimality will be discussed thereafter.

Proposition 7.2.1. *Let the assumption from Proposition 7.1.1 be fulfilled. Then, for the iterates $u^{(n)} = (\mathbf{u}^{(n)})^\top \Psi$ from Algorithm 17, it holds that*

$$\|u - u^{(n)}\|_{H^t(\Omega)} \leq \varepsilon_n.$$

Proof. The basic strategy for the proof is to carefully balance the error made in the solution of the subproblems, the application of the projector and the coarsening step with the error reduction of the exact Newton method. The assertion will be shown by an induction over n. For $n = 0$, the claim is valid by choice of ε_0. Let us assume now that the assertion is valid for some $n \geq 0$. In particular, this implies that $\|u^{(n)} - u\| \leq \varepsilon_n \leq \varepsilon_0$, thus $u^{(n)} \in U$. To show the inductive step, let $v^{(n)}$ be the exact Newton direction in $u^{(n)}$, and $\bar{u}^{(n+1)}$ the result of an exact Newton step in $u^{(n)}$, which means that

$$\bar{u}^{(n+1)} = u^{(n)} + v^{(n)} \tag{7.11}$$

with $D\mathcal{R}(u^{(n)})v^{(n)} = -\mathcal{R}(u^{(n)})$. Then, the convergence result from Proposition 7.1.1, $\|\bar{u}^{(n+1)} - u\|_{H^t(\Omega)} \leq \|u^{(n)} - u\|_{H^t(\Omega)}$, together with the norm constants C_E and M_2 gives

$$
\begin{aligned}
\|v^{(n)}\|_{u^{(n)}} &\leq \|\bar{u}^{(n+1)} - u\|_{u^{(n)}} + \|u^{(n)} - u\|_{u^{(n)}} \\
&\leq C_E M_2 (\|\bar{u}^{(n+1)} - u\|_{H^t(\Omega)} + \|u^{(n)} - u\|_{H^t(\Omega)}) \\
&\leq 2 C_E M_2 \|u^{(n)} - u\|_{H^t(\Omega)}.
\end{aligned}
$$

Now, by the induction hypothesis $\|u - u^{(n)}\|_{H^t(\Omega)} \leq \varepsilon_n$, we have

$$\|v^{(n)}\|_{u^{(n)}} \leq 2 C_E M_2 \varepsilon_n. \tag{7.12}$$

Recall that the inner iterates $v^{(n,k)}$ are an approximation to the $\bar{v}^{(n,k)}$ from Equation (7.7), calculated with the inexact solutions $\tilde{e}_i^{(n,k)}$ of the subproblems. Hence, from Lemma 7.1.3 and tolerances for the solution of the subproblems, we obtain

$$
\begin{aligned}
\|v^{(n,k+1)} - v^{(n)}\|_{u^{(n)}} &\leq \rho \|v^{(n,k)} - v^{(n)}\|_{u^{(n)}} + \omega \sum_{i=0}^{m-1} \|e_i^{(n,k)} - \tilde{e}_i^{(n,k)}\|_{u^{(n)}} \\
&\leq \rho \|v^{(n,k)} - v^{(n)}\|_{u^{(n)}} + (1 - \rho) C_E M_2 \varepsilon_n \bar{\rho}^k. \tag{7.13}
\end{aligned}
$$

With this estimate at hand, we show by an inner induction over k that

$$\|v^{(n,k)} - v^{(n)}\|_{u^{(n)}} \leq 2C_E M_2 \tilde{\rho}^k \varepsilon_n. \tag{7.14}$$

For $k = 0$, this assertion is true because of the bound (7.12) and $v^{(n,0)} = 0$. For the inductive step, combining (7.13) with the induction hypothesis (7.14) already gives

$$\|v^{(n,k+1)} - v^{(n)}\|_{u^{(n)}} \leq \rho \cdot 2C_E M_2 \tilde{\rho}^k \varepsilon_n + (1 - \rho)C_E M_2 \varepsilon_n \tilde{\rho}^k = 2\tilde{\rho}^{k+1} C_E M_2 \varepsilon_n,$$

which completes the inner induction. Hence, by the choice of k^*, we obtain

$$\|v^{(n,k^*)} - v^{(n)}\|_{u^{(n)}} \leq 2\tilde{\rho}^{k^*} C_E M_2 \varepsilon_n \leq \frac{1}{2} \frac{c_E}{M_1} \left(\frac{1}{2C^* + 1} \varepsilon_{n+1} - M_1 M_2 \beta \varepsilon_n^2 \right).$$

Thus, together with estimate (7.5), we infer

$$\|\mathbf{P}v^{(n,k^*)} - \mathbf{P}v^{(n)}\|_{\ell_2(\Lambda)} \leq \frac{1}{2M_1} \left(\frac{1}{2C^* + 1} \varepsilon_{n+1} - M_1 M_2 \beta \varepsilon_n^2 \right). \tag{7.15}$$

Now, let $\bar{u}^{(n+1)} = (\bar{\mathbf{u}}^{(n+1)})^\top \Psi$ from Equation (7.11) be the result of the exact Newton step starting from $u^{(n)}$, $\bar{\mathbf{u}}^{(n+1)} = \mathbf{u}^{(n)} + \mathbf{v}^{(n)}$. Then, with the tolerance with which $\hat{\mathbf{u}}^{(n+1)}$ is calculated, we have

$$\|\hat{\mathbf{u}}^{(n+1)} - \mathbf{P}\bar{\mathbf{u}}^{(n+1)}\|_{\ell_2(\Lambda)}$$
$$\leq \|\hat{\mathbf{u}}^{(n+1)} - \mathbf{P}(\mathbf{u}^{(n)} + \mathbf{v}^{(n,k^*)})\|_{\ell_2(\Lambda)} + \|\mathbf{P}(\mathbf{u}^{(n)} + \mathbf{v}^{(n,k^*)}) - \mathbf{P}(\mathbf{u}^{(n)} + \mathbf{v}^{(n)})\|_{\ell_2(\Lambda)}$$
$$\leq \frac{1}{2M_1} \left(\frac{1}{2C^* + 1} \varepsilon_{n+1} - M_1 M_2 \beta \varepsilon_n^2 \right) + \|\mathbf{P}(\mathbf{v}^{(n,k^*)} - \mathbf{v}^{(n)})\|_{\ell_2(\Lambda)}.$$

Combining this with the bound from (7.15) gives

$$\|\hat{\mathbf{u}}^{(n+1)} - \mathbf{P}\bar{\mathbf{u}}^{(n+1)}\|_{\ell_2(\Lambda)} \leq \frac{1}{M_1} \left(\frac{1}{2C^* + 1} \varepsilon_{n+1} - M_1 M_2 \beta \varepsilon_n^2 \right). \tag{7.16}$$

This inequality will be needed below. Furthermore, using the estimate (7.10), the error reduction (7.3) of the exact Newton iteration and the induction hypothesis $\|u^{(n)} - u\|_{H^t(\Omega)} \leq \varepsilon_n$, it holds that

$$\|\mathbf{P}\bar{\mathbf{u}}^{(n+1)} - \mathbf{P}u\|_{\ell_2(\Lambda)} \leq M_2 \|\bar{u}^{(n+1)} - u\|_{H^t(\Omega)} \leq M_2 \beta \|u^{(n)} - u\|_{H^t(\Omega)}^2 \leq M_2 \beta \varepsilon_n^2.$$

Together with the bound (7.16), we obtain

$$\|\hat{\mathbf{u}}^{(n+1)} - \mathbf{P}u\|_{\ell_2(\Lambda)} \leq \|\hat{\mathbf{u}}^{(n+1)} - \mathbf{P}\bar{\mathbf{u}}^{(n+1)}\|_{\ell_2(\Lambda)} + \|\mathbf{P}\bar{\mathbf{u}}^{(n+1)} - \mathbf{P}u\|_{\ell_2(\Lambda)}$$
$$\leq \frac{1}{M_1} \left(\frac{1}{2C^* + 1} \varepsilon_{n+1} - M_1 M_2 \beta \varepsilon_n^2 \right) + M_2 \beta \varepsilon_n^2$$
$$= \frac{1}{M_1} \frac{1}{2C^* + 1} \varepsilon_{n+1}. \tag{7.17}$$

Using the tolerance for the coarsening step, we have

$$\|\mathbf{u}^{(n+1)} - \mathbf{Pu}\|_{\ell_2(\Lambda)} \leq \|\mathbf{u}^{(n+1)} - \hat{\mathbf{u}}^{(n+1)}\|_{\ell_2(\Lambda)} + \|\hat{\mathbf{u}}^{(n+1)} - \mathbf{Pu}\|_{\ell_2(\Lambda)}$$
$$\leq \frac{1}{M_1} \frac{2C^*}{2C^*+1} \varepsilon_{n+1} + \frac{1}{M_1} \frac{1}{2C^*+1} \varepsilon_{n+1}$$
$$= \frac{1}{M_1} \varepsilon_{n+1}.$$

Finally, with the help of the inequality (7.9), we conclude

$$\|u^{(n+1)} - u\|_{H^t(\Omega)} \leq M_1 \|\mathbf{u}^{(n+1)} - \mathbf{Pu}\|_{\ell_2(\Lambda)} \leq M_1 \frac{1}{M_1} \varepsilon_{n+1} = \varepsilon_{n+1},$$

which completes the proof. $\qquad\square$

Besides from the convergence of the algorithm, we are also interested in the sparsity of its output. Similarly to Lemma 4.3.7, with the help of the properties of the coarsening method and the assumption on the boundedness of $\mathbf{P}$, we can already now conclude that the algorithm is asymptotically optimal with respect to the degrees of freedom.

Proposition 7.2.2. *Let the assumption from Proposition 7.1.1 be fulfilled and $\mathbf{P}$ be bounded on $\mathcal{A}_{\mathcal{AT}}^s$. Assume that we have a representation $u = \mathbf{u}^\top \Psi$ of the solution with $\mathbf{u} \in \mathcal{A}_{\mathcal{AT}}^s$. Then, all iterates $\mathbf{u}^{(n)}$, $n \geq 1$, from Algorithm 17 are contained in $\mathcal{A}_{\mathcal{AT}}^s$ with*

$$\|\mathbf{u}^{(n)}\|_{\mathcal{A}_{\mathcal{AT}}^s} \lesssim \|\mathbf{u}\|_{\mathcal{A}_{\mathcal{AT}}^s}$$

and

$$\# \operatorname{supp} \mathbf{u}^{(n)} \lesssim \varepsilon_n^{-1/s} \|\mathbf{u}\|_{\mathcal{A}_{\mathcal{AT}}^s}^{1/s}.$$

Proof. From the estimate (7.17), we have $\|\hat{\mathbf{u}}^{(n+1)} - \mathbf{Pu}\|_{\ell_2(\Lambda)} \leq \frac{1}{M_1} \frac{1}{2C^*+1} \varepsilon_{n+1}$. Then, from the Coarsening Lemma 4.2.3, we infer that $\|\mathbf{u}^{(n+1)}\|_{\mathcal{A}_{\mathcal{AT}}^s} \lesssim \|\mathbf{Pu}\|_{\mathcal{A}_{\mathcal{AT}}^s}$ and that $\# \operatorname{supp} \mathbf{u}^{(n+1)} \lesssim (\frac{1}{M_1} \varepsilon_{n+1})^{-1/s} \|\mathbf{Pu}\|_{\mathcal{A}_{\mathcal{AT}}^s}^{1/s}$. Together with the boundedness assumption on $\mathbf{P}$, this shows the assertion. $\qquad\square$

7.3 Remarks on the solution of the subproblems

In the previous section, we have left open the question of how to deal with the linear subproblems that need to be solved within Algorithm 17. These subproblems appear in a similar form as in the adaptive Schwarz methods in Chapter 4. Therefore, it is natural to take a similar approach for their numerical solution, compare Subsections 4.3.3 and 4.4.2. A main difference, however, is that the bilinear form $a_{u^{(n)}}(\cdot, \cdot)$ depends on the current global iterate $u^{(n)}$. This has significant consequences, not necessarily for the convergence theory, but more so for the practical realization. To explain this, we consider the discretized form of the local subproblems, which analogously to Subsection 4.4.2 reads as

$$(D\mathbf{R}(\mathbf{u}^{(n)})^{(i,i)})\mathbf{e}_i^{(n,k)} = -((D\mathbf{R}(\mathbf{u}^{(n)}))\mathbf{v}^{(n,k)})_{|\Lambda_i} - \mathbf{R}(\mathbf{u}^{(n)})_{|\Lambda_i}, \qquad (7.18)$$

where

$$D\mathbf{R}(\mathbf{v}) = \mathbf{A} + D\mathbf{G}(\mathbf{v}) = \mathbf{A} + (\langle G'(\mathbf{v}^\top\Psi)\psi_\mu, \psi_\lambda\rangle)_{\lambda,\mu\in\Lambda} \qquad (7.19)$$

is the discrete derivative of the residual and $(D\mathbf{R}(\mathbf{v}))^{(i,j)}$ is its (i,j)-th diagonal block, i.e., the part of the matrix belonging to the the the row indices from Λ_i and the column indices from Λ_j. Because $\mathbf{A}$ is positive semidefinite and G is monotone, $G'(x) \geq 0$, from Equation (7.19) we infer that $D\mathbf{R}(\mathbf{v})$ is positive semidefinite. Moreover, analogously to Lemma 2.3.3, we see that the main diagonal blocks $(D\mathbf{R}(\mathbf{v})^{(i,i)})$ are positive definite and boundedly invertible. Therefore, for the solution of the subproblem (7.18), we can in principle apply a Richardson iteration following the lines from Subsections 4.3.3 and 4.4.2. If we denote the exact right-hand side of the subproblems (7.18) by $\mathbf{r} := -((D\mathbf{R}(\mathbf{u}^{(n)}))\mathbf{v}^{(n,k)})_{|\Lambda_i} - \mathbf{R}(\mathbf{u}^{(n)})_{|\Lambda_i}$, for a sufficiently small relaxation parameter $\alpha > 0$, the exact Richardson iteration takes the form

$$\mathbf{y}^{(r+1)} = \mathbf{y}^{(r)} + \alpha(\mathbf{r} - (D\mathbf{R}(\mathbf{u}^{(n)})^{(i,i)})\mathbf{y}^{(r)})$$

with a given initial approximation such as $\mathbf{y}^{(0)} = \mathbf{0}$. Hence, an implementable version of this iteration requires the evaluation of matrix-vector products with the matrix $D\mathbf{R}(\mathbf{u}^{(n)})$ and its main diagonal blocks, at least up to given tolerances. A natural approach would be to try to evaluate this matrix-vector product with a scheme like **APPLY** as outlined in Subsection 4.2.2. This, however, cannot be achieved at acceptable computational costs, because already the computation of a single entry of the matrix $D\mathbf{G}(\mathbf{u}^{(n)})$, which is given by $\langle G'((\mathbf{u}^{(n)})^\top\Psi)\psi_\mu, \psi_\lambda\rangle$, requires the expensive approximation of a nonlinear function, compare also [27]. An alternative approach from loc. cit. for the evaluation of these matrix-vector products is to consider $(D\mathbf{R}(\mathbf{u}^{(n)}))\mathbf{v}$ or similar expressions with main diagonal blocks of $(D\mathbf{R}(\mathbf{u}^{(n)}))$ as nonlinear expressions in two variables. Then, an evaluation routine similar to the one described in Subsection 4.2.3 is applied. The basic idea is to first call the support prediction method separately for both arguments $\mathbf{u}^{(n)}$ and $\mathbf{v}$ of $(D\mathbf{R}(\mathbf{u}^{(n)}))\mathbf{v}$. Then, we approximately evaluate the nonlinear term on the union of both of these trees calculated by the support prediction step. However, this method has only been constructed so far for the case of wavelet bases. For the aggregated frames we work with in this thesis, such a construction has not yet been completely analyzed. Nevertheless, we expect that with techniques similar to those from [76], the generalization to wavelet frames, also for nonlinearities of several arguments, is possible. For the practical implementation of the method in the following subsection, we make use of the above idea and see that with this construction, the evaluation of these terms can be realized, albeit still at comparatively high computational costs.

7.4 Numerical tests

In this section, we test the Newton-Schwarz algorithm on the reference problem from Section 5.3, the weak form of the equation

$$-\Delta u + u^3 = f \text{ in } \Omega, \quad u = 0 \text{ on } \partial\Omega$$

on the L-shaped domain. For the practical realization of the method, the general remarks from Section 5.1 still apply similarly. In view of the experience from the previous tests, we make use of the spline wavelet basis from [95] with parameters $(l, \tilde{l}) = (3, 3)$. However, to achieve manageable run-times, in the case of the Newton-Schwarz method, we have to restrict the maximum level of wavelets considered to $j_{\max} = 6$, compare the discussion in Chapter 5. Therefore, the results in this section are not fully comparable to those from Section 5.3. The numerical experiments suggest that, from a computational point of view, it is not necessarily best to choose quadratically decreasing tolerances. This is because, in this case, even the tolerances for the local solver decrease quadratically. We have observed that in this configuration it tends to be more difficult to properly coordinate the tolerances of all the steps involved. Instead, when choosing linearly decreasing tolerances ε_n, $\varepsilon_{n+1} = \frac{4}{5}\varepsilon_n$, we can again make use of a constant number of iterations within the local solver and the method seems rather robust against changes in the estimates for the constants involved. We make a total of 35 iterations up to an approximate residual of around $5 \cdot 10^{-3}$.

In Figure 7.1, in the graph on the left-hand side, we show the degrees of freedom in terms of $\# \operatorname{supp} \mathbf{u}^{(n)}$ on the horizontal axis and the ℓ_2-norm of approximate discrete residual on the vertical axis, both in logarithmic scale. We compare the observed rate to the expected optimal rate, which, in view of the results from Chapter 3, compare also Chapter 5, is given by $\frac{l-t}{d} = \frac{3-1}{2} = 1$ and indicated by a reference line. We observe that, after an initial phase, the slope of the line produced by the algorithm is very close to that of the reference line. This result shows that the expected asymptotic behavior predicted by Proposition 7.2.2 is practically realized at relevant scales. On the right-hand side of the figure, we see a similar graph with the degrees of freedom replaced by the computational time. Similar to the degrees of freedom, at least in later stages of the iteration, even for this benchmark we observe a rate close to $\frac{l-t}{d} = \frac{3-1}{2} = 1$. This hints at a linear complexity in terms of the degrees of freedom, implying that the algorithm also shows asymptotically optimal behavior with respect to the computational effort, compare Definition 3.4.10. However, we have to concede that the constants involved, and thus the computational time in absolute terms, are rather high even compared to the algorithms from Chapter 4. This is mainly due to the costly evaluation of the matrix-vector products with the matrices $D\mathbf{R}(\mathbf{u}^{(n)})$, compare also the discussion in Section 7.3. At this point, there is probably room for practical improvements leading to a more efficient implementation.

In Figure 7.2, we see the local approximations $u_i = (\mathbf{u}_\varepsilon)^\top_{|\Lambda_i} \Psi^{(i)} \in H^t_0(\Omega_i)$, $i = 0, 1$ produced by the algorithm. Because, as in the implementation of the other Schwarz

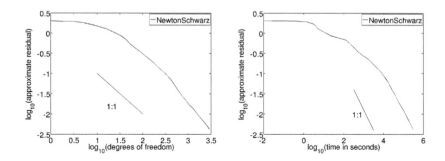

Figure 7.1: Degrees of freedom vs. ℓ_2-norm of the approximate residual (left) and computational time vs. ℓ_2-norm of the approximate residual (right) for the test of the Newton-Schwarz method on the L-shaped domain.

methods discussed in Chapter 5, we leave out the explicit application of the projector $\mathbf{P}$, it not clear how the approximate solution $u_\varepsilon = u_0 + u_1$ decomposes into these local parts. We observe in the figure that this representation does not seem to be very redundant. Hence, leaving out the projection step in practice seems to be justified even for the Newton-Schwarz method.

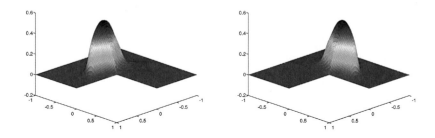

Figure 7.2: Local approximations of the solution from the Newton-Schwarz algorithm on Ω_0 (left) and Ω_1 (right).

In Figure 7.3, we show the pointwise error at the end of the iteration. This error takes its maximum around the singularity of the solution at the origin with a value of around $8 \cdot 10^{-3}$. Even if, for the Newton-Schwarz method, we only consider wavelets up to $j_{\max} = 6$, this is of a similar magnitude than the results for the multiplicative and additive Schwarz methods from Chapter 5.

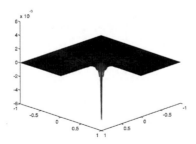

Figure 7.3: Pointwise error in the Newton-Schwarz method.

To sum up, we have seen that with the tools and techniques we have at hand, a Newton-Schwarz method can be realized and that the expected asymptotic behavior can practically be observed. Nevertheless, the performance seems to lack somewhat behind the results for the previously discussed Schwarz methods. This is due to the more complex evaluation of the matrices $D\mathbf{R}(\mathbf{u}^{(n)})$ involved in the algorithm. Nevertheless, a Newton-Schwarz method might show a larger potential to be generalized to different kinds of nonlinear equations. Therefore, it seems worthwhile to further study this kind of methods.

Conclusion

In this thesis, we have developed adaptive wavelet Schwarz methods for a range of nonlinear elliptic partial differential equations. These methods were theoretically analyzed, practically implemented and tested. We have seen that the construction principles can furthermore be adapted to the stationary, incompressible Navier-Stokes equation. In addition, we have formulated a combination of a Schwarz method with a Newton iteration.

Let us discuss the results in a little more detail with respect to the tasks (T1) to (T4) formulated in the introduction and give perspectives for further work. The first task (T1), the development and analysis of adaptive wavelet methods for semilinear problems, was addressed in Chapter 4. The main results are the theorems 4.3.10 and 4.4.3. These results state that, at least for suitable nonlinearities that fulfill a contraction property, the multiplicative and the additive method are convergent. In addition, under a technical assumption, it is proven that they are asymptotically optimal in the sense that they reproduce the convergence rate of the best N-term tree approximation and show linear complexity in the degrees of freedom. Thus, task (T1) can, to a large extent, be considered fulfilled.

The practical implementation and the test of the methods, i.e., the solution of task (T2), was addressed in Chapter 5. In experiments in one and two space dimensions, we have seen that the theoretically expected optimal convergence rates of the algorithms are observed in practice. Furthermore, we have seen that the adaptive methods outperform a comparable standard uniform method. Hence, task (T2) is solved as well. Nevertheless, an improvement of the absolute runtimes would be desirable.

In Chapter 6, we have addressed task (T3) and developed an adaptive wavelet method for the stationary, incompressible Navier-Stokes equation. Applying divergence-free wavelets, we have seen that the techniques applied for the solution of task (T1) could be adapted to this task. The central results of this chapter were Theorem 6.3.1, showing the convergence of the method for sufficiently small Reynolds numbers, and Proposition 6.3.2, its optimality with respect to the degrees of freedom. Thus, we have largely solved task (T3). However, a practical implementation of this method has not yet been done.

The last remaining task (T4), the combination of adaptive wavelet Schwarz methods with Newton's method, was addressed in Chapter 7. There, we have developed an adaptive wavelet Newton-Schwarz method and verified its convergence for semilinear elliptic equations with monotone and stable nonlinearities. We have shown its convergence and verified that it is asymptotically optimal with respect to the degrees of freedom. In numerical experiments on the L-shaped domain, we have seen that the optimal convergence rate is realized in practice. The quantitative performance of this method, however, is not yet fully satisfactory. As well as for the Navier-Stokes equation, an estimate for the computational effort would also be desirable to achieve a complete solution to task (T4).

Regarding all of the tasks treated in this thesis, there are perspectives for future work. Within all the methods presented, one could consider the application of sparsening strategies, that in [111] were applied to reduce the degrees of freedom in the overlapping region and made it possible to show optimality of an adaptive multiplicative Schwarz method for *linear* problems without the need for a projection step as in Subsection 4.3.1. However, these strategies seem to strongly rely on the linearity of the equation. Nevertheless, it would be desirable to avoid the application of this projector and the assumption on its boundedness on the tree approximation space $\mathcal{A}^s_{\mathcal{AT}}$. Regarding the additive and multiplicative Schwarz method, another interesting question is whether the convergence theory can be generalized to other kinds of nonlinearities, for which the theory from [92] might serve as a starting point. A large potential for the generalization to a broader range of nonlinear problems seems to lie in the study of the Newton-Schwarz method. To this end, the analysis of the application of Newton's method to partial differential equations from [27, 51] might be helpful. Regarding the method for the stationary, incompressible Navier-Stokes equation, it would be desirable to establish convergence for larger Reynolds numbers. This, however, probably requires a different approach than the linearization strategy we have chosen, because for large Reynolds numbers, the nonlinear term becomes dominant.

Concerning the results for the Navier-Stokes equation and the Newton-Schwarz method, for a full analysis of the complexity of the local solvers, it would furthermore be necessary to study the generalization of the evaluation scheme for the Jacobi matrices that arise in Newton's method and the quadratic nonlinearity from the Navier-Stokes equation. This has not yet been done, but we expect this to be possible with the techniques from [27], which, however, only cover wavelet Riesz bases. A generalization to aggregated wavelet frames is so far known for Nemitsky operators of one argument, see [75]. Spending effort on the study evaluation schemes for nonlinear functions in wavelet coordinates also seems an important point from a practical point of view. This is because with currently known schemes, a major part of the computational costs is required for exactly this step. In addition, to further improve the runtime of the algorithms, it would be desirable to study the algorithmic aspect of the involved building blocks. Furthermore, the generalization of the implementation to problems in three space dimensions should be considered.

As a related research project, it is intended within the workgroup *Numerics and Optimization* at Philipps-Universität Marburg to study the construction of interpolating wavelets on bounded domains. On the long run, these wavelets could give rise to adaptive wavelet methods with fast evaluation schemes. Furthermore, it would be important to study to which extent the construction principles of the overlapping domain decomposition methods presented here carry over to other kinds of equations.

For instance, adaptive wavelet methods for integral equations have made significant progress in recent years, see, e.g., [71], and might be a field of application for Schwarz methods. Another promising development is the construction of adaptive wavelet schemes using anisotropic wavelets, since they allow for convergence rates independent of the space dimension, see, e.g. [55]. A combination of such methods with overlapping domain decomposition solvers could be considered in order to benefit from the advantages of both approaches.

List of Figures

Bibliography

[1] *Wavelet and Multiscale Library*, `http://www.mathematik.uni-marburg.de/` `~waveletsoft/`, Accessed: 2014-07-24.

[2] R.A. Adams and J.J.F. Fournier, *Sobolev spaces*, Pure and Applied Mathematics, Elsevier Science, 2003.

[3] H.-B. An, *On convergence of the additive Schwarz preconditioned inexact Newton method*, SIAM J. Numer. Anal. **43** (2005), no. 5, 1850–1871.

[4] J Arnal, V Migallón, J Penadés, and D-B. Szyld, *Newton additive and multiplicative Schwarz iterative methods*, IMA J. Numer. Anal. **28** (2008), no. 1, 143–161.

[5] M. Badiale and E. Serra, *Semilinear elliptic equations for beginners*, Universitext, Springer, London, 2011, Existence results via the variational approach.

[6] A. Barinka, *Fast computation tools for adaptive wavelet schemes*, PhD thesis, RWTH Aachen, 2005.

[7] A. Barinka, W. Dahmen, and R. Schneider, *Fast computation of adaptive wavelet expansions*, Numer. Math. **105** (2007), no. 4, 549–589.

[8] J. Bergh and J. Löfström, *Interpolation spaces. An introduction*, Springer-Verlag, Berlin-New York, 1976, Grundlehren der Mathematischen Wissenschaften, No. 223.

[9] P. Binev, W. Dahmen, and R. A. DeVore, *Adaptive finite element methods with convergence rates*, Numer. Math. **97** (2004), no. 2, 219–268.

[10] P. Binev and R. A. DeVore, *Fast computation in adaptive tree approximation*, Numer. Math. **97** (2004), no. 2, 193–217.

[11] J. H. Bramble, *Multigrid methods*, Pitman Research Notes in Mathematics Series, vol. 294, Longman Scientific & Technical, Harlow; copublished in the United States with John Wiley & Sons, Inc., New York, 1993.

[12] X.-C. Cai and M. Dryja, *Domain decomposition methods for monotone nonlinear elliptic problems*, Domain decomposition methods in scientific and engineering computing (University Park, PA, 1993), Contemp. Math., vol. 180, Amer. Math. Soc., Providence, RI, 1994, pp. 21–27.

[13] X.-C. Cai and David E. Keyes, *Nonlinearly preconditioned inexact Newton algorithms*, SIAM J. Sci. Comput. **24** (2002), no. 1, 183–200.

[14] C. Canuto, A. Tabacco, and K. Urban, *The wavelet element method. I. Construction and analysis*, Appl. Comput. Harmon. Anal. **6** (1999), no. 1, 1–52.

[15] _____, *The wavelet element method. II. Realization and additional features in 2D and 3D*, Appl. Comput. Harmon. Anal. **8** (2000), no. 2, 123–165.

[16] J. M. Carnicer, W. Dahmen, and J. M. Peña, *Local decomposition of refinable spaces and wavelets*, Appl. Comput. Harmon. Anal. **3** (1996), no. 2, 127–153.

[17] M. Charina, C. Conti, and M. Fornasier, *Adaptive frame methods for nonlinear variational problems*, Numer. Math. **109** (2008), no. 1, 45–75.

[18] O. Christensen, *An introduction to frames and Riesz bases*, Applied and Numerical Harmonic Analysis, Birkhäuser, Boston, 2003.

[19] O. Christensen and K. L. Christensen, *Approximation theory. From Taylor polynomials to wavelets.*, Birkhäuser, Boston, 2004.

[20] C. K. Chui, *An introduction to wavelets*, Academic Press Professional, Inc., San Diego, CA, USA, 1992.

[21] P. A. Cioica, S. Dahlke, N. Döhring, S. Kinzel, F. Lindner, T. Raasch, K. Ritter, and R. L. Schilling, *Adaptive wavelet methods for the stochastic Poisson equation*, BIT **52** (2012), no. 3, 589–614.

[22] P. A. Cioica, S. Dahlke, S. Kinzel, F. Lindner, T. Raasch, K. Ritter, and R. L. Schilling, *Spatial Besov regularity for stochastic partial differential equations on Lipschitz domains*, Studia Math. **207** (2011), no. 3, 197–234.

[23] A. Cohen, *Numerical analysis of wavelet methods*, Studies in Mathematics and its Applications, vol. 32, North-Holland Publishing Co., Amsterdam, 2003.

[24] A. Cohen, W. Dahmen, I. Daubechies, and R. A. DeVore, *Tree approximation and optimal encoding*, Appl. Comput. Harmon. Anal. **11** (2001), no. 2, 192–226.

[25] A. Cohen, W. Dahmen, and R. A. DeVore, *Adaptive wavelet methods for elliptic operator equations: convergence rates*, Math. Comp. **70** (2001), no. 233, 27–75.

[26] _____, *Adaptive wavelet methods. II. Beyond the elliptic case*, Found. Comput. Math. **2** (2002), no. 3, 203–245.

[27] _____, *Adaptive wavelet schemes for nonlinear variational problems*, SIAM J. Numer. Anal. **41** (2003), no. 5, 1785–1823.

[28] _____, *Sparse evaluation of compositions of functions using multiscale expansions*, SIAM J. Math. Anal. **35** (2003), no. 2, 279–303.

[29] A. Cohen, I. Daubechies, and J.-C. Feauveau, *Biorthogonal bases of compactly supported wavelets*, Comm. Pure Appl. Math. **45** (1992), no. 5, 485–560.

[30] S. Dahlke, *Besov regularity for elliptic boundary value problems in polygonal domains*, Appl. Math. Lett. **12** (1999), no. 6, 31–36.

[31] S. Dahlke, W. Dahmen, and K. Urban, *Adaptive wavelet methods for saddle point problems—optimal convergence rates*, SIAM J. Numer. Anal. **40** (2002), no. 4, 1230–1262.

[32] S. Dahlke and R. A. DeVore, *Besov regularity for elliptic boundary value problems*, Comm. Partial Differential Equations **22** (1997), no. 1-2, 1–16.

[33] S. Dahlke, M. Fornasier, M. Primbs, T. Raasch, and M. Werner, *Nonlinear and adaptive frame approximation schemes for elliptic PDEs: theory and numerical experiments*, Numer. Methods Partial Differential Equations **25** (2009), no. 6, 1366–1401.

[34] S. Dahlke, M. Fornasier, and T. Raasch, *Adaptive frame methods for elliptic operator equations*, Adv. Comput. Math. **27** (2007), no. 1, 27–63.

[35] S. Dahlke, M. Fornasier, T. Raasch, R. Stevenson, and M. Werner, *Adaptive frame methods for elliptic operator equations: the steepest descent approach*, IMA J. Numer. Anal. **27** (2007), no. 4, 717–740.

[36] S. Dahlke, R. Hochmuth, and K. Urban, *Adaptive wavelet methods for saddle point problems*, M2AN Math. Model. Numer. Anal. **34** (2000), no. 5, 1003–1022.

[37] _____, *Convergent adaptive wavelet methods for the Stokes problem*, Multigrid methods, VI (Gent, 1999), Lect. Notes Comput. Sci. Eng., vol. 14, Springer, Berlin, 2000, pp. 66–72.

[38] S. Dahlke, D. Lellek, S. H. Lui, and R. Stevenson, *Adaptive wavelet Schwarz methods for the Navier-Stokes equation*, DFG-SPP 1324 Preprint series (2013), no. 140.

[39] S. Dahlke and W. Sickel, *On Besov regularity of solutions to nonlinear elliptic partial differential equations*, Rev. Mat. Complut. **26** (2013), no. 1, 115–145.

[40] W. Dahmen, *Stability of multiscale transformations*, J. Fourier Anal. Appl. **2** (1996), no. 4, 341–361.

[41] _____, *Wavelet and multiscale methods for operator equations*, Acta numerica, 1997, Acta Numer., vol. 6, Cambridge Univ. Press, Cambridge, 1997, pp. 55–228.

[42] _____, *Wavelet and multiscale methods for operator equations*, Acta Numerica **6** (1997), 55–228.

[43] W. Dahmen, H. Harbrecht, and R. Schneider, *Adaptive methods for boundary integral equations: complexity and convergence estimates*, Math. Comp. **76** (2007), no. 259, 1243–1274.

[44] W. Dahmen, A. Kunoth, and K. Urban, *Biorthogonal spline wavelets on the interval—stability and moment conditions*, Appl. Comput. Harmon. Anal. **6** (1999), no. 2, 132–196.

[45] W. Dahmen and R. Schneider, *Wavelets with complementary boundary conditions—function spaces on the cube*, Results Math. **34** (1998), no. 3-4, 255–293.

[46] _____, *Composite wavelet bases for operator equations*, Math. Comp. **68** (1999), no. 228, 1533–1567.

[47] W. Dahmen, R. Schneider, and Y. Xu, *Nonlinear functionals of wavelet expansions—adaptive reconstruction and fast evaluation*, Numer. Math. **86** (2000), no. 1, 49–101.

[48] I. Daubechies, *Ten lectures on wavelets*, CBMS-NSF Regional Conference Series in Applied Mathematics, vol. 61, Society for Industrial and Applied Mathematics (SIAM), Philadelphia, PA, 1992.

[49] F. Demengel and G. Demengel, *Functional spaces for the theory of elliptic partial differential equations*, Universitext, Springer, London; EDP Sciences, Les Ulis, 2012.

[50] E. Deriaz and V. Perrier, *Direct numerical simulation of turbulence using divergence-free wavelets.*, Multiscale Model. Simul. **7** (2008), no. 3, 1101–1129.

[51] P. Deuflhard, *Newton methods for nonlinear problems*, Springer Series in Computational Mathematics, vol. 35, Springer, Berlin, 2004, Affine invariance and adaptive algorithms.

[52] R. A. DeVore, *Nonlinear approximation*, Acta numerica, 1998, Acta Numer., vol. 7, Cambridge Univ. Press, Cambridge, 1998, pp. 51–150.

[53] R. A. DeVore and G. G. Lorentz, *Constructive approximation*, Grundlehren der Mathematischen Wissenschaften, vol. 303, Springer, Berlin, 1993.

[54] R. A. DeVore and V. A. Popov, *Interpolation of Besov spaces*, Trans. Amer. Math. Soc. **305** (1988), no. 1, 397–414.

[55] T. J. Dijkema, C. Schwab, and R. Stevenson, *An adaptive wavelet method for solving high-dimensional elliptic PDEs*, Constr. Approx. **30** (2009), no. 3, 423–455.

[56] S. Dispa, *Intrinsic characterizations of Besov spaces on Lipschitz domains*, Math. Nachr. **260** (2003), 21–33.

[57] M. Dobrowolski, *Angewandte Funktionalanalysis. Funktionalanalysis, Sobolev-Räume und elliptische Differentialgleichungen.*, Springer, Berlin, 2006.

[58] M. Dryja and W. Hackbusch, *On the nonlinear domain decomposition method*, BIT **37** (1997), no. 2, 296–311.

[59] M. Dryja and O. Widlund, *An additive variant of the Schwarz alternating method for the case of many subregions*, New York University, Courant Institute of Mathematical Sciences, Computer Science Technical Report (1987), no. 339.

[60] _____, *Towards a unified theory of domain decomposition algorithms for elliptic problems*, Third International Symposium on Domain Decomposition Methods for Partial Differential Equations (Houston, TX, 1989), SIAM, Philadelphia, PA, 1990, pp. 3–21.

[61] F. Eckhardt, *Besov regularity for the stokes and the navier-stokes system in polyhedral domains*, ZAMM - Journal of Applied Mathematics and Mechanics / Zeitschrift für Angewandte Mathematik und Mechanik (2014).

[62] L. C. Evans, *Partial differential equations*, Graduate Studies in Mathematics, vol. 19, American Mathematical Society, Providence, RI, 1998.

[63] M. J. Gander, *Schwarz methods over the course of time*, Electron. Trans. Numer. Anal. **31** (2008), 228–255.

[64] T. Gantumur and R. Stevenson, *Computation of differential operators in wavelet coordinates*, Math. Comp. **75** (2006), no. 254, 697–709.

[65] V. Girault and P.-A. Raviart, *Finite element approximation of the Navier-Stokes equations*, Lecture Notes in Mathematics, vol. 749, Springer, Berlin, 1979.

[66] P. Grisvard, *Elliptic problems in nonsmooth domains*, Monographs and Studies in Mathematics, vol. 24, Pitman (Advanced Publishing Program), Boston, MA, 1985.

[67] _____, *Singularities in boundary value problems*, Recherches en Mathématiques Appliquées, vol. 22, Masson, Paris, 1992.

[68] W. Hackbusch, *Elliptic differential equations – Theory and numerical treatment*, Springer Series in Computational Mathematics, vol. 18, Springer, Berlin, 1992.

[69] _____, *Multi-grid methods and applications*, Springer Series in Computational Mathematics, Springer, Berlin, Heidelberg, 2003.

[70] H. Harbrecht and R. Schneider, *Wavelet Galerkin schemes for boundary integral equations—implementation and quadrature*, SIAM J. Sci. Comput. **27** (2006), no. 4, 1347–1370 (electronic). MR 2199752 (2006j:65379)

[71] _____, *Rapid solution of boundary integral equations by wavelet Galerkin schemes*, Multiscale, nonlinear and adaptive approximation, Springer, Berlin, 2009, pp. 249–294.

[72] Q. He and D. J. Evans, *Monotone-schwarz parallel algorithm for nonlinear elliptic equations*, Parallel Algorithms Appl. **9** (1996), no. 3-4, 173–184.

[73] D. Jerison and C. E. Kenig, *The inhomogeneous Dirichlet problem in Lipschitz domains*, J. Funct. Anal. **130** (1995), no. 1, 161–219.

[74] Y. Jiang, *Divergence-free wavelet solution to the Stokes problem*, Anal. Theory Appl. **23** (2007), no. 1, 83–91.

[75] J. Kappei, *Adaptive frame methods for nonlinear elliptic problems*, Appl. Anal. **90** (2011), no. 8, 1323–1353.

[76] _____, *Adaptive wavelet frame methods for nonlinear elliptic problems*, Logos Verlag Berlin, 2011.

[77] C.T. Kelley, *Solving nonlinear equations with Newton's method.*, Philadelphia, PA: SIAM Society for Industrial and Applied Mathematics, 2003.

[78] M. Kovács, S. Larsson, and K. Urban, *On wavelet-Galerkin methods for semi-linear parabolic equations with additive noise*, Monte Carlo and quasi-Monte Carlo methods 2012, Springer Proc. Math. Stat., vol. 65, Springer, Heidelberg, 2013, pp. 481–499.

[79] E. Kreyszig, *Introductory functional analysis with applications*, Wiley Classics Library, John Wiley & Sons, Inc., New York, 1989.

[80] S. Larsson and V. Thomée, *Partial differential equations with numerical methods*, Texts in Applied Mathematics, vol. 45, Springer, Berlin, 2009, Paperback reprint of the 2003 edition.

[81] D. Lellek, *Adaptive Frame-Verfahren für elliptische Operatorgleichungen: Verfeinerungen und neue Strategien*, Diploma Thesis, Philipps-Universität Marburg, 2010.

[82] _____ , *Adaptive wavelet frame domain decomposition methods for nonlinear elliptic equations*, Numer. Methods Partial Differential Equations **29** (2013), no. 1, 297–319.

[83] P. G. Lemarie-Rieusset, *Analyses multi-résolutions non orthogonales, commutation entre projecteurs et dérivation et ondelettes vecteurs à divergence nulle*, Rev. Mat. Iberoamericana **8** (1992), no. 2, 221–237.

[84] P.-L. Lions, *On the Schwarz alternating method. I*, First International Symposium on Domain Decomposition Methods for Partial Differential Equations (Paris, 1987), SIAM, Philadelphia, PA, 1988, pp. 1–42.

[85] S. H. Lui, *On Schwarz alternating methods for nonlinear elliptic PDEs*, SIAM J. Sci. Comput. **21** (1999/00), no. 4, 1506–1523.

[86] _____ , *On Schwarz alternating methods for the incompressible Navier-Stokes equations*, SIAM J. Sci. Comput. **22** (2000), no. 6, 1974–1986.

[87] _____ , *On linear monotone iteration and Schwarz methods for nonlinear elliptic PDEs*, Numer. Math. **93** (2002), no. 1, 109–129.

[88] _____ , *Numerical analysis of partial differential equations*, Pure and Applied Mathematics (Hoboken), John Wiley & Sons Inc., Hoboken, NJ, 2011.

[89] S. G. Mallat, *Multiresolution approximations and wavelet orthonormal bases of $L_2(\mathbb{R})$*, Trans. Amer. Math. Soc. **315** (1989), no. 1, 69–87.

[90] Y. Meyer, *Ondelettes et opérateurs. I*, Actualités Mathématiques., Hermann, Paris, 1990, Ondelettes.

[91] P. Morin, R. H. Nochetto, and K. G. Siebert, *Convergence of adaptive finite element methods.*, SIAM Rev. **44** (2002), no. 4, 631–658 (English).

[92] M. Munteanu and L. F. Pavarino, *An overlapping additive Schwarz-Richardson method for monotone nonlinear parabolic problems*, Domain decomposition methods in science and engineering XVII, Lect. Notes Comput. Sci. Eng., vol. 60, Springer, Berlin, 2008, pp. 599–606.

[93] M.J.D. Powell, *Approximation theory and methods*, Cambridge University Press, 1981.

[94] M. Primbs, *Stabile biortogonale Spline-Waveletbasen auf dem Intervall*, PhD thesis, Universität Duisburg-Essen, 2006.

[95] _____ , *New stable biorthogonal spline-wavelets on the interval*, Results Math. **57** (2010), no. 1-2, 121–162.

[96] A. Quarteroni and A. Valli, *Domain decomposition methods for partial differential equations*, Numerical Mathematics and Scientific Computation, The Clarendon Press, Oxford University Press, New York, 1999, Oxford Science Publications.

[97] T. Raasch, *Adaptive wavelet and frame schemes for elliptic and parabolic equations*, Logos Verlag Berlin, 2007.

[98] M. Renardy and R. C. Rogers, *An introduction to partial differential equations*, second ed., Texts in Applied Mathematics, vol. 13, Springer, New York, 2004.

[99] T. Runst and W. Sickel, *Sobolev spaces of fractional order, Nemytskij operators, and nonlinear partial differential equations*, de Gruyter Series in Nonlinear Analysis and Applications, vol. 3, Walter de Gruyter & Co., Berlin, 1996.

[100] Y. Saad, *Iterative methods for sparse linear systems*, second ed., Society for Industrial and Applied Mathematics, Philadelphia, PA, 2003.

[101] K. Schneider, M. Farge, F. Koster, and M. Griebel, *Adaptive wavelet methods for the Navier-Stokes equations*, Numerical flow simulation, II, Notes Numer. Fluid Mech., vol. 75, Springer, Berlin, 2001, pp. 303–318.

[102] C. Schwab, *p- and hp-finite element methods*, Numerical Mathematics and Scientific Computation, The Clarendon Press, Oxford University Press, New York, 1998, Theory and applications in solid and fluid mechanics.

[103] H. A. Schwarz, *Gesammelte mathematische Abhandlungen*, vol. 2, Springer, Berlin, 1890.

[104] R. Stevenson, *Adaptive solution of operator equations using wavelet frames*, SIAM J. Numer. Anal. **41** (2003), no. 3, 1074–1100.

[105] _____, *On the compressibility operators in wavelet coordinates*, SIAM J. Math. Anal. **35** (2004), no. 5, 1110–1132.

[106] _____, *Optimality of a standard adaptive finite element method*, Found. Comput. Math. **7** (2007), no. 2, 245–269.

[107] _____, *Adaptive wavelet methods for solving operator equations: an overview*, Multiscale, nonlinear and adaptive approximation, Springer, Berlin, 2009, pp. 543–597.

[108] _____, *Divergence-free wavelet bases on the hypercube: free-slip boundary conditions, and applications for solving the instationary Stokes equations*, Math. Comp. **80** (2011), no. 275, 1499–1523.

[109] _____, *Divergence-free wavelets on the hypercube: General boundary conditions*, ESI preprint (2013), no. 2417.

[110] R. Stevenson and M. Werner, *Computation of differential operators in aggregated wavelet frame coordinates*, IMA J. Numer. Anal. **28** (2008), no. 2, 354–381.

[111] _____, *A multiplicative Schwarz apdative wavelet method for elliptic boundary value problems*, Math. Comp. **78** (2009), no. 266, 619–644.

[112] X.-C. Tai and M. Espedal, *Applications of a space decomposition method to linear and nonlinear elliptic problems*, Numer. Methods Partial Differential Equations **14** (1998), no. 6, 717–737.

[113] X.-C. Tai and J. Xu, *Global and uniform convergence of subspace correction methods for some convex optimization problems*, Math. Comp. **71** (2002), no. 237, 105–124.

[114] R. Temam, *Navier-Stokes equations. Theory and numerical analysis*, North-Holland Publishing Co., Amsterdam, 1977, Studies in Mathematics and its Applications, Vol. 2.

[115] A. Toselli and O. Widlund, *Domain decomposition methods—algorithms and theory*, Springer Series in Computational Mathematics, vol. 34, Springer, Berlin, 2005.

[116] H. Triebel, *Theory of function spaces*, Monographs in Mathematics, vol. 78, Birkhäuser, Basel, 1983.

[117] _____, *Theory of function spaces. II*, Monographs in Mathematics, vol. 84, Birkhäuser, Basel, 1992.

[118] _____, *Theory of function spaces. III*, Monographs in Mathematics, vol. 100, Birkhäuser, Basel, 2006.

[119] K. Urban, *Wavelets in numerical simulation*, Lecture Notes in Computational Science and Engineering, vol. 22, Springer, Berlin, 2002, Problem adapted construction and applications.

[120] R. Verfürth, *A posteriori error estimation techniques for nonlinear elliptic and parabolic pde's.*, Rev. Eur. Élém. Finis **9** (2000), no. 4, 377–402.

[121] J. Vorloeper, *Adaptive Wavelet Methoden für Operator Gleichungen Quantitative Analyse und Softwarekonzepte*, vol. 20, VDI Fortschritt Berichte, no. 427, VDI Verlag, 2010.

[122] M. Werner, *Adaptive Frame-Verfahren für Elliptische Randwertprobleme*, Diploma Thesis, Philipps-Universität Marburg, 2005.

[123] _____ , *Adaptive wavelet frame domain decomposition methods for elliptic operator equations*, Logos Verlag Berlin, 2009.

[124] B. I. Wohlmuth, *Discretization methods and iterative solvers based on domain decomposition.*, Springer, Berlin, 2001 (English).

[125] P. Wojtaszczyk, *A mathematical introduction to wavelets*, London Mathematical Society Student Texts, vol. 37, Cambridge University Press, Cambridge, 1997.

[126] J. Xu, *Iterative methods by space decomposition and subspace correction*, SIAM Rev. **34** (1992), no. 4, 581–613.

[127] Y. Xu and Q. Zou, *Adaptive wavelet methods for elliptic operator equations with nonlinear terms*, Adv. Comput. Math. **19** (2003), no. 1-3, 99–146, Challenges in computational mathematics (Pohang, 2001).